The Laue Method

THE LAUE METHOD

José Luis Amorós

School of Engineering and Technology
Southern Illinois University
Carbondale, Illinois

Martin J. Buerger

Department of Earth and Planetary Science
Massachusetts Institute of Technology
Cambridge, Massachusetts
    and
Institute of Materials Science
University of Connecticut
Storrs, Connecticut

Marisa Canut de Amorós

School of Engineering and Technology
Southern Illinois University
Carbondale, Illinois

1975

ACADEMIC PRESS

New York   San Francisco   London

A Subsidiary of Harcourt Brace Jovanovich, Publishers

ACADEMIC PRESS, INC.
111 Fifth Avenue, New York, New York 10003

United Kingdom Edition published by
ACADEMIC PRESS, INC. (LONDON) LTD.
24/28 Oval Road, London NW1

Library of Congress Cataloging in Publication Data

Amorós, José Luis.
    The Laue method.

    Includes bibliographies.
    1.   X-ray crystallography.   I.   Buerger, Martin
Julian, Date   joint author.   II.   Canut de Amorós,
M. L., joint author.   III.   Title.
QD945.A458   548'.83   73-5311
ISBN 0–12–057450–0

Contents

Chapter 4   The stereognomonic projection

Chapter 5   The conventional Laue method

Chapter 6   Symmetry determination with the aid of Laue photographs

Chapter 7   Crystallochemical analysis

Chapter 8 Planes and zones

Chapter 9 The polychromatic component

Chapter 10 The cross ratio and its application in crystallography

Chapter 11 The indexing of Laue photographs

Chapter 12 The characteristic component of Laue photographs and its application to the study of diffuse scattering

Chapter 13 The optics of the Laue method

Chapter 14 Epilogue: Graphical interpretation of the Laue method

Preface

The Laue method was the vehicle for the discovery of the diffraction of x rays by crystals sixty years ago. It marked the beginning of a tremendous advance in science which led to our present detailed knowledge of the structure of crystalline matter. We enjoyed the advantages of a modern technology of solid materials in no small measure as a direct consequence of this discovery.

The Laue method was the forerunner of other ways of using x-ray diffraction for the study of matter. To each of these methods there are characteristic advantages and disadvantages. Because it has its own advantages, the Laue method is still utilized, at least to some extent, in substantially every x-ray diffraction laboratory. Unfortunately, the treatment of the Laue method is now confined to, at most, a chapter in books which deal with the entire field of x-ray diffraction, and, indeed, the only book devoted to this venerable method, Schiebold's "Die Lauemethode," is now forty years old and no longer available. To discuss the Laue method properly and to make known some of its advantages, this book was written.

The fantastic pace at which the science of x-ray crystallography has advanced has left the Laue method in a half-forgotten corner, basically because it is not well suited to the popular field of crystal-structure analysis. Its commonest use is to check and adjust the orientation of a crystal which is destined to be used in another method, probably leading to the determination of its crystal structure. The Laue method is also used to assess the perfection of a crystal, although this application is not used by all who might well benefit from it. One of its more recent and important applications is in the identification and study of disorder in crystals by the analysis of diffuse scattering. For this purpose, a relatively rapid survey of all reciprocal space is required, for which the Laue method offers special advantages. Many who might well study the disorder in crystals do not do so because the application of the Laue method in this connection is not generally recognized and understood.

The use of the Laue method in studying the cell and symmetry of a crystal is practiced by relatively few scientists, and then chiefly in the study of the simplest crystals. We have taken the trouble to discuss just what can be ascertained in this field, and how the investigation can best be done. It is a curious fact that, before the Laue method was discovered, an elegant body of crystallographic calculus had been developed. This was applied to getting crystallographically useful data from the angular measurements made on the surface planes of crystals with the aid of the optical goniometer. While the calculus applied to such data did yield some useful results, the data were too limited to lead to the results really desired. For example, because the original data were so limited, the attempt to identify crystals by using the calculus to rework the data was so close to a failure that only a few devotees attempted to identify crystals from such data. The situation actually changed drastically when the Laue method was discovered, for this supplied an overabundance of data. It took some sixty years to realize this, however, for the attention of crystallographers had turned almost completely to the solution of easy crystal structures, and to finding a general solution for all crystal structures. Accordingly, we have taken the trouble to demonstrate why and how the Laue method provides an easy vehicle for identification of crystalline species.

The Laue method should have an appeal to various kinds of scientists who study crystals. It is obvious from what has just been said that the method supplies a natural research vehicle for those interested in the calculus and projection methods of classical crystallography. Accordingly, the more important aspects of classical crystal theory and projection methods (such as the stereographic, gnomonic and stereognomonic projections) are carefully developed in this book. It is our hope that those interested in classical crystallography will discover that the older crystallography acquires a new significance in terms of the interpretation of crystal data available in the Laue method, which furnishes a speedy and simultaneous goniometry of all crystal planes of interest. We also hope that this easy goniometry and the abundant information that the photographs contain may encourage many metallurgists, who ordinarily make only a limited use of the information available, to take advantage of the additional information which these photographs could still provide.

On another level, the dual nature of the information contained in the Laue photograph should always be kept in mind. The sharp spots of the photograph provide information about the reciprocal-lattice points contained in the volume of the Ewald sphere defined by the minimum wavelength, while the background provides information, through the monochromatic component, concerning the part of the reciprocal space on the surface of the Ewald sphere corresponding to the characteristic radiation.

The subject matter of this book falls naturally into two parts. The first part, consisting of Chapters 2 through 8, provides, at an elementary level, a simple and compact treatment of the Laue method and the background needed to make use of it. The second part, Chapters 9 through 13, treat the Laue method on a higher level; this is written for the research scientist who would like to exploit all the unique advantages inherent in the Laue method. In addition to these two ranges of chapters which constitute the main subject matter of the book, the book begins with a Prologue, in which the originating ideas, first experiment, and initial interpretation are analyzed in historical retrospect; the book ends with an Epilogue which is concerned with a very simple new interpretation of the Laue method.

This book is accordingly written for both student and specialist. Both will find the Laue method a useful technique for the study of crystals. The references have been selected so that the basic points of each chapter are supported by significant literature so that the reader can deepen his knowledge and extend his study into related fields.

We wish especially to express our appreciation to Dr. L. V. Azároff, Director of the Institute of Materials Science at the University of Connecticut, for providing appropriate facilities (including a temporary residence) so that the coauthors could bring their respective contributions to the chapters into mutually acceptable form.

Chapter 1

Prologue

Over 50 years have elapsed since Laue made his historic discovery of x-ray diffraction by crystals. Although the story of this turning point in science has often been told, it is especially appropriate to use it as an introduction to a book on the Laue method.

Historical background

About 1912, an exceptional group of scientists was in residence in Munich; among others were Professor Paul von Groth (the dean of crystallographers), Professor Wilhelm Konrad Röntgen (the discoverer of x rays), and Professor A. Sommerfeld (the well-known theoretical physicist). Associated with Röntgen were a number of people working in experimental research on the nature and properties of x rays: Von Angerer was measuring the energy of such radiation; Bassler, its polarization; and Friedrich, the azimuthal distribution of its intensity. In the field of theoretical physics, the numerous papers on the theory of x rays by Sommerfeld, a follower of Boltzmann, were well known. Max von Laue joined Sommerfeld's group in the fall of 1909. He was a pupil of Planck and had obtained his degree in Berlin. At that time, he was working on the theory of interference and wave optics.

Of no less importance was the presence of the old master of crystallography, Paul von Groth, whose views on the molecular structure of

1

crystals were well known. For many years crystallographers had realized that crystals were discontinuous solids with three-dimensional repetitive order, and it was believed that the motifs of their patterns were molecules. This idea, however, was little more than a good working hypothesis. It was thought that the crystal faces were expressions of the lattice aspect of the pattern while cleavage was an experimental evidence of the discontinuous aspect which would be expected because of the lattice translations. The close connection of crystallographers and physicists therefore provided an appropriate setting for the discovery which was to follow.

The starting spark for the discovery was the doctoral thesis of P. P. Ewald, in which he attempted to account for the optical properties of crystals as an interaction of atomic dipoles with the electromagnetic waves of visible light. Ewald showed his thesis to Laue a few days before submitting it to the Philosophical Faculty on 16 February 1912. Ewald explained to Laue that in his study of the dispersion he had assumed the resonators to be situated at lattice points, because crystals were thought by crystallographers to have such internal regularity. Laue then asked what was the distance between resonators and what would happen if very much shorter waves would travel through the crystal. Ewald pointed out that the derivation of one of the equations of his manuscript thesis was valid also for short wavelengths.

From a consideration of specific gravity, molecular weight, and the mass of the hydrogen atom, the lattice translations of crystals were known to be of the order of 10^{-8} cm. Moreover, Wien and Sommerfeld had shown that the wavelength of the x rays should be of the order of 10^{-9} cm. Diffraction by a three-dimensional grating had never been considered, but, as Laue pointed out later on [†], his optical intuition told him at that time that if the wavelength is of the same magnitude as the atomic distances in the regular arrangement in the crystal, this must lead to some kind of diffraction effect with these shorter wavelengths.

Laue discussed his feelings with Sommerfeld, Wien, and others. It was argued that the temperature motions of the atoms would disrupt the regularity of the grating to such an extent that no pronounced maxima could be expected.

Laue further discussed the matter from a theoretical point of view in a seminar; the opinion then prevailed that experiment was safer than theory. As a consequence of his enthusiasm and clarity of ideas, Friedrich, at that time Sommerfeld's assistant, became interested in the problem, as did Paul Knipping, a research student who had just finished his thesis work in

[†] Max von Laue, On the discovery of x-ray interference. Nobel Prize Lecture, Stockholm (June 3, 1920).

Röntgen's institute. Both volunteered to assist in an experimental test of this theory. Friedrich and Knipping then developed the provisional experimental arrangement schematically shown in Fig. 1, which is a reproduction of their original drawing. An x-ray beam of 1-mm cross section was isolated by means of four lead screens B_1 to B_4 from the hemisphere of radiation produced by the anticathode A of a Muller x-ray bulb. The x-ray beam so selected passed through a crystal Kr mounted on a goniometer head G. The x-ray bulb had been provided by Röntgen and the crystal by von Groth. A crystal of copper sulfate was used as the diffraction grating, and in different directions and at different distances, there were placed photographic plates on which the diffracted beams were to be recorded. In order to prevent the undesired direct radiation from reaching the photographic plate, a larger planar lead screen S and lead case K were placed between the x-ray bulb and other parts of the apparatus. The tube was operated with a current of 2–10 mA and the exposure time varied between 1 and 12 hours. Initial pitfalls and the subsequent improvement of the technique can be deduced from Fig. 2, which contains copies of the

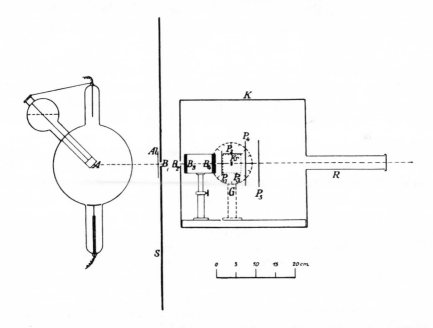

Fig. 1. Scheme of the original setup of Friedrich and Knipping's experiment. A: anticathode; Al, aluminum foil; S, lead screen; B_1, B_2, B_3, B_4, openings of the collimator system; Kr, crystal; G, pedestal; P_1, P_2, P_3, P_4, P_5, photographic plates; K, lead case; R, beam stop. [From Friedrich *et al.*[1], p. 313.]

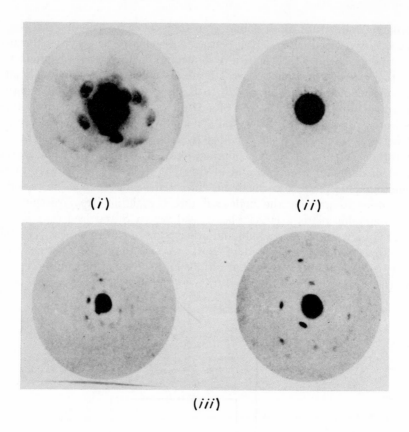

(*i*)   (*ii*)

(*iii*)

Fig. 2. Reproduction of the original Laue photographs of Friedrich and Knipping. (*i*) The first picture; (*ii*) picture obtained from a powder sample; (*iii*) other attempts. [From Friedrich *et al.*[1], Tafeln I and II.]

original photographs from the research by Friedrich, Knipping, and Laue that was communicated to the Bavarian Academy of Sciences at the meetings of 8 June and 6 July 1912 by A. Sommerfeld, a fellow of the academy.

Friedrich and Knipping, both true experimentalists, took every precaution to reassure themselves, as well as possible critics, that the observed phenomenon was due to the presence of crystalline material. Photographs taken with the beam irradiating different portions of the same crystal gave constant results. Friedrich and Knipping showed a great deal of naivety, even taking a photograph of the pulverized crystal to show that no diffraction effect was observed from a noncrystalline material (Fig. 2*ii*).

The analysis of the photographs obtained from the irradiated single crystal showed the presence of spots other than the one produced by the primary x-ray beam. This, together with fact that by moving the photographic plate backward or forward it could be seen that the spots were formed by rectilinear pencils spreading in all directions from the crystal, showed the experimenters that diffraction really did occur, so that Laue's intuition was confirmed. Thus, there was experimental proof both of the periodic nature of the crystals and of the wave nature of x rays. At the same time the *Laue method* was born.

Early interpretation of Laue photographs

The initial interpretation of the experimental results was due to Laue. This was one of those rare occasions on which one knows the exact place and moment of the discovery. Laue himself gives a vivid narration of the event.[4]

> It was not the first, but the second [picture] that gave a result. The [x-ray] transmission photograph of a piece of copper sulfate showed near the primary x-ray beam a crown of diffracted lattice spectra. Deep in thought, I was heading home through Leopoldstrasse after Friedrich had showed to me the photograph. And very near to my home, Bismarkstrasse 22, by the house Sigfriedstrasse 10, occurred to me the idea for the mathematical theory of the result.

Shortly before that day, in fact, Laue had written an article for the *Enzyklopaedie der mathematischen Wissenshaften* in which he had given a new foundation to the old theory of diffraction by an optical grating: by applying the equation of the theory twice over, the theory of diffraction by a cross grating could be obtained. Laue realized then that in order to interpret the new discovery, he had to write the equation three times, one for each periodicity of the space lattice. In only a week's work, he was able to have a quantitative theory of the diffraction of x rays by crystals. Thus, the famous Laue equations were formulated.

The theory was based on the application of monochromatic x rays, as required by interference theory. At that time Laue, Friedrich, and Knipping were convinced that the diffracted rays should consist of characteristic radiation emitted by the crystal under the influence of the incident ray. The choice of the copper sulfate crystal was made with this in mind. Thus Laue tried to associate the diffraction spots of the diagram with five different wavelengths. Because he was unable to get exact results, he thought that his equations held only approximately.

The correct interpretation came a few months later, not from his group, but from another researcher, William Lawrence Bragg, a young student of physics at Cambridge and the son of William Henry Bragg, Cavendish Professor of Physics at the University of Leeds. It was not by chance that the Braggs became interested at once in Laue's experiment and theory. The father was already experienced with x rays, having set up the first x-ray tube in Adelaide (Australia) shortly after Röntgen's discovery. At the time of Laue's discovery, W. H. Bragg held the theory that x rays were a type of corpuscular radiation. In fact, in October 1912 he proposed an alternative explanation of the fourfold symmetry of the Laue photograph of sphalerite; he proposed that all the directions of the secondary pencils in this position of the crystal are "avenues" between the atoms of the crystal. His son was actually making some unsuccessful experiments to get evidence of his father's views. They soon, however, accepted the wave theory of x rays as explaining the diffraction experiments. Nevertheless, W. L. Bragg was convinced that Laue's analysis of the x-ray photograph was not correct. Instead of the small number of wavelengths assumed by Laue, Bragg proposed the existence of a continuous spectrum in the incident radiation. At the same time, he considered the crystal as a diffraction grating from a point of view different from that of Laue, and this led him to important simplifications. Bragg showed that the spots in Laue's photographs could be explained as partial reflection of the incident beam in sets of parallel planes on which the atoms were arranged in the crystal. On trying the Laue experiment with a sheet of mica, Bragg proved that the laws of reflection were obeyed, and he was able to formulate the condition for diffraction in a very simple and compact form, the famous Bragg equation. Bragg also showed that Laue's equations were satisfied not approximately, but rather rigorously. He showed that exposures of only a few minutes, instead of long hours, were sufficient to take a Laue photograph.

At Cambridge, interaction between physicists and crystallographers proved to be successful. By studying Pope's and Barlow's papers, W. L. Bragg became familiar with their views on crystal structure, and he was able to demonstrate that the Laue photograph of sphalerite was characteristic of a face-centered cubic crystal. Pope further encouraged W. L. Bragg to try Laue photographs of NaCl and KCl, for which Barlow had long since proposed what is now known as the sodium chloride structure. These analyses established the structure of the sodium chloride group of crystals and crystal-structure determination was initiated.

Laue's theory of diffraction by crystals

A monochromatic wave of wavelength λ and unit intensity can be represented by

$$z_1 = \exp[ik(ct - \alpha_0 x - \beta_0 y - \gamma_0 z)] \tag{1}$$

where c is the velocity of light; k a constant equal to $2\pi/\lambda$; t the time; α_0, β_0, γ_0 the direction cosines of the wave normal in the orthogonal system x, y, z. The planes

$$x\alpha_0 + y\beta_0 + z\gamma_0 = \text{constant} \tag{2}$$

are the planes of equal phase of that wave. If the time term is disregarded, (1) can be written as

$$z_1 = \exp[-ik(\alpha_0 x + \beta_0 y + \gamma_0 z)]. \tag{3}$$

Let us assume now that this monochromatic wave impinges on a monoatomic crystal in which the positions of the atoms are given by

$$\mathbf{r}_l = l_1\mathbf{a}_1 + l_2\mathbf{a}_2 + l_3\mathbf{a}_3 \tag{4}$$

where $\mathbf{a}_i$ are the crystallographic translation vectors. Under the influence of the incident wave, an atom P_1 becomes the origin of a secondary spherical wave that, at a distance R_1 from the atom, can be expressed as a periodic function, such as

$$z_2 = (\psi/R_1) \exp(-ikR_1) \tag{5}$$

where ψ is the amplitude diffracted by the atom. These new waves have the same wavelength as the incident wave because the diffraction is coherent. Each atom in the crystal thus becomes a secondary center of emission. The resulting amplitude from the whole crystal is just the product of (1) and (5), so that

$$A = \sum_{l_1} \sum_{l_2} \sum_{l_3} (\psi/R_1) \exp[-ik(R_1 + \alpha_0 x_1 + \beta_0 y_1 + \gamma_0 z_1)] \tag{6}$$

where x_1, y_1, z_1, are the orthogonal coordinates of the atom at the lattice point $P(l_1 l_2 l_3)$.

The interference effect of the wavelets generated by the various atoms is observed at a distance R_0 from the center of the crystal. Because this distance is very large in relation to the dimensions of the crystal, it can be assumed that $R_0 = R_1$, in which case the amplitude at this point is given by

$$A'_{R_0} = \psi' \frac{\exp(-ikR_0)}{R_0}$$

$$\times \sum_{l_1} \sum_{l_2} \sum_{l_3} \exp\{ik[(\alpha - \alpha_0)x_1 + (\beta - \beta_0)y_1 + (\gamma - \gamma_0)z_1]\} \quad (7)$$

where α, β, γ are the direction cosines of the direction R_0; at the same time we know that

$$x_1 = l_1 a_{1x} + l_2 a_{2x} + l_3 a_{3x},$$

$$y_1 = l_1 a_{1y} + l_2 a_{2y} + l_3 a_{3y}, \quad (8)$$

$$z_1 = l_1 a_{1z} + l_2 a_{2z} + l_3 a_{3z}$$

where the a_{ix}, a_{iy}, a_{iz} $(i = 1, 2, 3)$ are the components of the base vectors $\mathbf{a}_1$, $\mathbf{a}_2$, $\mathbf{a}_3$ on the orthogonal reference system. Laue introduced the reduction

$$A_1 = k[(\alpha - \alpha_0)a_{1x} + (\beta - \beta_0)a_{1y} + (\gamma - \gamma_0)a_{1z}],$$

$$A_2 = k[(\alpha - \alpha_0)a_{2x} + (\beta - \beta_0)a_{2y} + (\gamma - \gamma_0)a_{2z}], \quad (9)$$

$$A_3 = k[(\alpha - \alpha_0)a_{3x} + (\beta - \beta_0)a_{3y} + (\gamma - \gamma_0)a_{3z}],$$

which allowed him to write (7) in the form

$$A_{R_0} = \psi \frac{\exp(-ikR_0)}{R_0}$$

$$\times \sum_{l_1=0}^{2M_1-1} \exp(iA_1 l_1) \sum_{l_2=0}^{2M_2-1} \exp(iA_2 l_2) \sum_{l_3=0}^{2M_3-1} \exp(iA_3 l_3). \quad (10)$$

Taking into account that

$$\sum_{l=0}^{2M_1-1} \exp(iA_1 l_1) = \frac{\sin M_1 A_1}{\sin \frac{1}{2}A_1} \exp\left[\frac{i(2M_1 - 1)A_1}{2}\right], \quad (11)$$

Eq. (10) can be written in a more condensed form,

$$A_{R_0} = \psi \frac{\exp(-ikR_0)}{R_0} \exp\{i[M_1 A_1 + M_2 A_2 + M_3 A_3 - \frac{1}{2}(A_1 + A_2 + A_3)]\}$$

$$\times \frac{\sin M_1 A_1}{\sin \frac{1}{2}A_1} \frac{\sin M_2 A_2}{\sin \frac{1}{2}A_2} \frac{\sin M_3 A_3}{\sin \frac{1}{2}A_3}. \quad (12)$$

The intensity at the point of observation is the product of A_{R_0} by its complex conjugate, so that finally Laue obtained the relation

$$I = \frac{|\psi|^2}{R_0^2} \left\{ \frac{\sin^2 M_1 A_1}{\sin^2 \frac{1}{2} A_1} \frac{\sin^2 M_2 A_2}{\sin^2 \frac{1}{2} A_2} \frac{\sin^2 M_3 A_3}{\sin^2 \frac{1}{2} A_3} \right\}. \tag{13}$$

The expression in the braces is the interference function; it has a maximum when

$$A_1 = 2\pi h_1, \qquad A_2 = 2\pi h_2, \qquad A_3 = 2\pi h_3. \tag{14}$$

By introducing these values into (13), Laue deduced that the intensity at a maximum is given by

$$I_{\max} = \frac{|\psi|^2}{R_0^2} (2M_1)^2 (2M_2)^2 (2M_3)^2. \tag{15}$$

A monoatomic crystal with translations a_1, a_2, a_3 contains $2M_1 \times 2M_2 \times 2M_3$ atoms, and therefore the intensities for all the maxima are equal and proportional to the square of the number of atoms. In nonmonoatomic crystals the relative coordinates of the atoms in the unit cell must be taken into account.

The direction of a maximum cannot be deduced from these equations. Laue, however, gave this direction in an explicit and condensed form:

$$(\alpha - \alpha_0) a_{1x} + (\beta - \beta_0) a_{1y} + (\gamma - \gamma_0) a_{1z} = h_1 \lambda,$$

$$(\alpha - \alpha_0) a_{2x} + (\beta - \beta_0) a_{2y} + (\gamma - \gamma_0) a_{2z} = h_2 \lambda, \tag{16}$$

$$(\alpha - \alpha_0) a_{3x} + (\beta - \beta_0) a_{3y} + (\gamma - \gamma_0) a_{3z} = h_3 \lambda.$$

These so-called *Laue equations* are best known in vectorial form:

$$\mathbf{a}_1 \cdot (\mathbf{s} - \mathbf{s}_0) = h_1 \lambda, \qquad \mathbf{a}_2 \cdot (\mathbf{s} - \mathbf{s}_0) = h_2 \lambda, \qquad \mathbf{a}_3 \cdot (\mathbf{s} - \mathbf{s}_0) = h_3 \lambda. \tag{17}$$

These expressions can be easily obtained from (16) by noting that

$$\alpha a_{1x} + \beta a_{1y} + \gamma a_{1z} = (\mathbf{s} \mathbf{a}_1)$$

$$\alpha_0 a_{1x} + \beta_0 a_{1y} + \gamma_0 a_{1z} = (\mathbf{s}_0 \mathbf{a}_1)$$

$$\vdots$$

The three integers h_1, h_2, h_3, known as the Laue numbers, give the orders of the interference.

The Laue equations (17) have a simple physical interpretation, due also to Laue. Each one of the equations (17) represents the diffraction produced by a single row of atoms with an interatomic interval such as a_1 (Fig. 3). Because the individual scattered waves are spherical, the effect of their superposition is the same in all planes containing the row of atoms. The direction $\mathbf{s}$ then represents rather a cone surrounding the

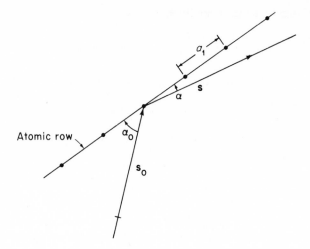

Fig. 3. Laue model of the diffraction by a row of atoms.

row; the maximum amplitude is obtained, in fact, in all directions which are generators of this cone. A pair of equations is equivalent to the diffraction effect due to a two-dimensional grating; the common solution of two equations implies that the maxima of the individual equations are the same only at the intersection of the cones produced by the two rows of atoms. The three equations of (17), finally, represent the diffraction by the three noncoplanar rows of the crystal, and the maxima are located along the intersections common to cones of the three cone systems.

Deduction of the Bragg equation from the Laue equations

Laue considered diffraction produced by strictly monochromatic x radiation. However, it can be easily deduced that if shorter wavelengths are also contained in the spectrum of x radiation impinging on the crystal, multiple-order diffraction maxima utilizing wavelengths $\lambda/2, \lambda/3, \ldots$ also occur in the same direction as the diffraction due to λ. This accounts for the multiplicity of the wavelengths in an individual spot of a Laue photograph. This possibility was correctly assumed by Bragg.

The famous Bragg equation, which assumes that the diffraction of x rays can be explained as a discontinuous reflection of the radiation incident on crystal planes, can be deduced from the Laue equations in the following way. Following Ewald, let us define a vector $\mathbf{r}^*$ (Fig. 4) such that

$$\mathbf{s} - \mathbf{s}_0 = \mathbf{r}^*. \tag{18}$$

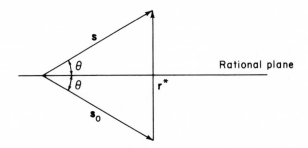

Fig. 4. Ewald conception of the Bragg model.

In this case the Laue equations are written as

$$\mathbf{a}_1 \cdot \mathbf{r}^* = h_1\lambda, \qquad \mathbf{a}_2 \cdot \mathbf{r}^* = h_2\lambda, \qquad \mathbf{a}_3 \cdot \mathbf{r}^* = h_3\lambda. \tag{19}$$

The implicit condition for (18) is that the three vectors $\mathbf{s}$, $\mathbf{s}_0$, and $\mathbf{r}^*$ are coplanar. If the vectors $\mathbf{s}$ (parallel to the diffraction beam) and $\mathbf{s}_0$ (parallel to the incident beam) are unit vectors, $\mathbf{r}^*$ is perpendicular to a plane that bisects the angle $\mathbf{s} \wedge \mathbf{s}_0$. Let us assume that this plane is a crystal plane. The intercepts of this crystal plane on the crystal axes a_1, a_2, a_3 must be integral numbers of translations from the origin. If OQ (Fig. 5) is the distance from the center of the crystal to the intersection of $\mathbf{r}^*$ with this plane, then

$$\frac{a_{1h}}{OQ} = \frac{|\mathbf{a}_1|}{|\mathbf{a}_1|/h}, \qquad \frac{a_{2h}}{OQ} = \frac{|\mathbf{a}_2|}{|\mathbf{a}_2|/k}, \qquad \frac{a_{3h}}{OQ} = \frac{|\mathbf{a}_3|}{|\mathbf{a}_3|/l} \tag{20}$$

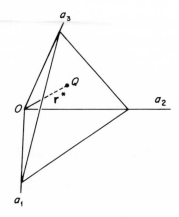

Fig. 5. A crystal plane and its distance to the origin.

where

$$a_{1h} = \frac{h_1\lambda}{|\mathbf{r}^*|}, \qquad a_{2h} = \frac{h_2\lambda}{|\mathbf{r}^*|}, \qquad a_{3h} = \frac{h_3\lambda}{|\mathbf{r}^*|} \qquad (21)$$

are the projections of $\mathbf{a}_1$, $\mathbf{a}_2$, $\mathbf{a}_3$ on $\mathbf{r}^*$, and h, k, l are the indices of the plane. Now we can deduce the following relations between the interference numbers h_1, h_2, h_3 and crystal-plane indices hkl:

$$h_1 = \frac{OQ}{\lambda}|\mathbf{r}^*|h, \qquad h_2 = \frac{OQ}{\lambda}|\mathbf{r}^*|k, \qquad h_3 = \frac{OQ}{\lambda}|\mathbf{r}^*|l. \qquad (22)$$

In other words $h_1{:}h_2{:}h_3 = h{:}k{:}l$. By definition h, k, l are coprimes, and therefore

$$\frac{OQ}{\lambda}|\mathbf{r}^*|h = n \qquad (23)$$

is an integer. We can take OQ equal to d, the spacing of the planes (hkl); in this case OQ is the smallest possible distance of the plane from the origin. Then (23) can be written as

$$\frac{d}{\lambda}|\mathbf{r}^*| = n. \qquad (24)$$

If we call 2θ the angle $\mathbf{s} \wedge \mathbf{s}_0$, it is obvious from the condition (18) that

$$\frac{|\mathbf{r}^*|}{2} = |\mathbf{s}_0|\sin\theta = |\mathbf{s}|\sin\theta. \qquad (25)$$

But $|\mathbf{s}_0| = |\mathbf{s}| = 1$, as previously stated; therefore

$$|\mathbf{r}^*| = 2\sin\theta. \qquad (26)$$

By substituting (26) in (24), we obtain

$$\frac{d}{\lambda}2\sin\theta = n \qquad (27)$$

or

$$2d\sin\theta = n\lambda, \qquad (28)$$

the famous *Bragg equation.*

Therefore, the diffraction of x rays by a crystal can be described either in terms of the diffraction by a three-dimensional array of atoms whose lattice translations are $\mathbf{a}_1$, $\mathbf{a}_2$, $\mathbf{a}_3$, or as a selective reflection of the x rays by the stack of planes (hkl) whose spacing is d. Both descriptions are valid, but the Bragg model is simpler and more straightforward. In the discussion of Laue photographs we shall make use of the Bragg model.

Importance, peculiarities, and applications of the Laue method

The Laue method is of special interest because historically it was the first of many methods developed for the study of x-ray diffraction by crystals. Many of the early determinations of cell dimensions and crystal symmetry and some early studies of crystal structure were made using this method. These uses showed the potentialities of the method. However, they also demonstrated the intrinsic limitations of the method when applied to crystal-structure determination. These limitations arise from the difficulty of knowing the exact wavelength and intensity of the part of the incident radiation that caused the observed Laue spot. It is only natural, then, that the development of more-powerful methods, which made use of moving-film instruments or diffractometers, determined a shift of interest from the old Laue method to newer techniques. Thus, the Laue method was reduced to an aid in determining crystal orientation and crystal perfection, although it is also used as a routine in the preliminary testing of crystal to be studied by more appropriate x-ray diffraction techniques. The method is still widely used in metallography when it is necessary to analyze crystals or determine orientation of highly absorbent materials. This use alone, however, would hardly justify a monograph like the present one if other advantages of unique character were not also offered by the Laue technique.

One of the principal reasons for studying the Laue method is that it is different from any other x-ray diffraction method. The Laue method utilizes the whole available spectrum of x-rays produced by the target of the x-ray tube rather than just the characteristic radiation used by the other x-ray diffraction methods. Due to this, peculiarities exist that are found only in the Laue method. For instance, the different orders of diffractions of a given crystal plane coincide in a single Laue spot. Accordingly, the reflections that are recorded in a Laue photograph are customarily labeled with indices that are relatively prime to one another, although most spots also have contributions from reflections with multiples of these indices. Thus, there is a one-to-one correspondence between a Laue spot and a rational plane of the crystal. For this reason a Laue photograph provides data for the goniometry of the rational planes of the crystal, so that the Laue method provides the link between classical (morphological) crystallography and x-ray crystallography. Accordingly, it is convenient to reintroduce the classical crystallographic calculus into modern crystallography.

In all x-ray diffraction methods the recording of the diffraction is a spatial function of θ, the Bragg glancing angle. In all other x-ray diffraction methods there is a one-to-one correspondence between θ and indices of the diffraction; therefore, the temperature factor, which is a function of sin θ, affects different regions of the film differently. The influence of the temperature factor has a greater effect on the reflections due to planes with complicated indices. The number and intensity of the reflections in a zone depend on the temperature factor, but this influence is not dependent on the region of the film on which the effect is observed.

Because the intensity of the characteristic radiation is higher by far than that of the continuous spectrum, the background modulation of the film is due to the characteristic radiation alone. In crystal-structure determination, in general, only the Bragg intensities are necessary; for this purpose the concept of a discrete diffraction space is useful. When dealing with the study of the disorder in crystals and crystal dynamics, however, the concept of a continuous diffraction space is more appropriate. In this case, the Laue method presents enormous advantages over other methods since it allows complete areas to be registered, and thus permits a complete survey of diffraction space. It constitutes, therefore, a powerful aid in the study of disorder of any kind in a crystal. Accordingly, the Laue method, especially what we call the systematic Laue survey of a crystal, is significant since today the importance of the study of diffuse scattering and its revelation of any kind of disorder in crystals is now recognized.

The fact that the crystal is maintained in a fixed orientation during the whole experiment makes the Laue method very sensitive to any kind of imperfection that may be present in the crystal. It therefore gives indication of mosaic structure and imperfections due to slip, and it becomes a useful tool in selecting crystals which are free of such imperfections, and which are needed for crystal-structure determination.

Finally, since the Laue method furnishes data for the goniometry of the rational crystal planes, it provides a direct way of determining both the orientation and symmetry of the crystal. Twinning can also be detected with Laue photographs and the twin laws determined. Furthermore, preferred orientation of grains in a polycrystalline sample can be studied via a systematic use of the Laue method along the sample.

Moreover, except for Schiebold's unavailable 1932 book, an occasional book chapter, and some brief papers, there are no general treatments of the Laue method. It is because of this void that we have written the present monograph.

The geometry of the Laue method is concerned with the way the diffracted beams, which radiate from the crystal as a center into three-dimensional space, are related to the primary beam. It is convenient to treat these relations by one of the standard devices for projecting three dimensions onto a two-dimensional surface. Accordingly, to lay the groundwork for this treatment of Laue photographs, the stereographic and gnomonic projections, as well as the stereognomonic projection, are discussed in advance of the discussion of the theory of interpreting Laue photographs, which constitutes the goal of this monograph.

Significant literature

1. W. Friedrich, P. Knipping, and M. Laue. Interferenz-Erscheinungen bei Röntgen-strahlen. *Sitzungsberichte der mathematisch-physikalischen Klasse der Königlich Bayerischen Akademie der Wissenschaften zu München*, 1912, 303–322; reprinted in *Naturwiss.* (1952) 361–367.
2. M. Laue. Eine quantitative Prüfung der Theorie für die Interferenz-Erscheinungen bei Röntgenstrahlen. *Sitzungsberichte der mathematisch-physikalischen Klasse der Königlich Bayerischen Akademie der Wissenschaften zu München*, 1912, 363–373: reprinted in *Naturwiss.* (1952) 368–372.
3. M. v. Laue. Röntgenstrahlinterferenzen. *Physik. Z.* 14 (1913) 1075–1079.
4. P. P. Ewald. Fifty years of x-ray diffraction. (N.V.A. Oosteroek's Vitgeversmaats-chappj, Utrecht, 1962) 720 pages, especially 31–56, 82–101.

Chapter 2

The stereographic projection

Projections in general

The word *projection* has a number of meanings; in the sense used here it is a geometrical scheme which permits representing certain three-dimensional relations in two dimensions. There are many such projections in common use. A well-known one is the orthographic projection, in which the features of a three-dimensional object are projected by parallel lines to a plane perpendicular to the lines; this is the way the object looks from a distant viewpoint (ideally, from infinity). A familiar cartographic projection is the Mercator projection, for which the features on the earth's surface are projected by rays from the earth's center to a cylindrical surface tangent to the earth's equator. The unrolled cylindrical surface constitutes the Mercator projection of the earth's surface.

Crystallographic problems are usually concerned with the relations between various directions and planes in three dimensions. The angular relations between these elements can be divorced from extraneous properties (such as the positions of the planes, as distinct from their orientations) by moving any line or plane parallel to itself until it contains the center of a sphere, and then mapping the intersection of the line or plane on the sphere's surface, as seen in Fig. 1. Such a representation of lines and planes is called a *spherical projection* of them. The sphere is usually called a *reference sphere*. Note that this sphere is neither the sphere of reflection of Chapter 5 nor the reflection sphere of Chapter 9.

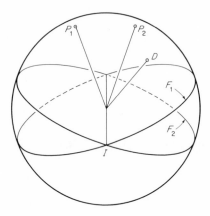

Fig. 1. Spherical projection of two planes F_1 and F_2, their intersection I, their normals P_1 and P_2, and an arbitrary direction D.

The reference sphere has a surface that cannot be flattened out on a sheet of paper, but it is possible to make various projections of it on a plane. A number of such projections have useful geometrical properties and can be used to aid the solution of crystallographic problems.

Perspective projections

When the eye, placed at a particular point in space, looks toward a three-dimensional object, the eye sees that object as if projected on a plane normal to the line of sight. Any projection of an object which has this geometry is called a *perspective projection*. In the application considered here, the various perspective projections of the reference sphere are of interest. The appearance of the sphere depends on the point from which the eye views it. Consider a point P on a sphere, Fig. 2, whose distance from the north pole N is defined by the angle ρ. Suppose the eye looks at the sphere from the side of the south pole. If the eyepoint is infinitely distant, point P appears projected in a plane tangent to the sphere at N, at a distance $R \sin \rho$ from the north pole; this eyepoint gives rise to the *orthographic projection* of the sphere. If the eye is placed at the south pole, the point P appears to be at a distance $2R \tan \frac{1}{2}\rho$ from N; this eyepoint gives a *stereographic projection* of the sphere. If the eye is placed at the center of the sphere, the point P appears to be at a distance $R \tan \rho$ from N; this eyepoint gives rise to the *gnomonic projection* of the sphere.

Each of these perspective projections has its own useful set of properties The stereographic and gnomonic projections have been used extensively in crystallography. The first is considered in this chapter, the other in the next chapter. While it is convenient to take the plane tangent to the sphere at N as the plane of the gnomonic projection, it is traditional to utilize the equatorial plane as the plane of the stereographic projection. The change in location of the plane of projection from that shown in Fig. 2 does not affect the properties of the stereographic projection other than to scale down the central distance of the projection of P to half value.

The angular relations for those perspective projections that can be made on the equatorial plane can be derived from Fig. 3. If the eye is placed at E, which is located at a distance a from the center of the sphere whose radius is taken as unity, the point P on the sphere's surface subtends an angle α from the eyepoint at E, and appears projected on (intercepted by) the equatorial plane at p. The distance s of point p from the center O of the projection is given by

$$\tan \alpha = \frac{s}{a} \tag{1}$$

If the law of sines is applied to triangle EOP, it is seen that

$$\frac{\sin \alpha}{1} = \frac{\sin(\rho - \alpha)}{a}, \tag{2}$$

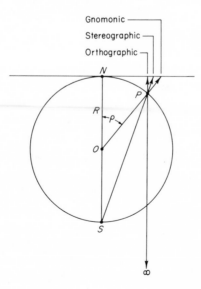

Gnomonic

Stereographic

Orthographic

Fig. 2. Relation of the projections of a point P on several kinds of prospective projections.

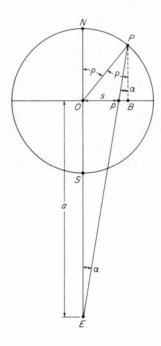

Fig. 3. Central distance *s* in a perspective projection on the equatorial plane of the reference sphere.

so that

$$a = \frac{\sin(\rho - \alpha)}{\sin \alpha} = \frac{\sin \rho \cos \alpha - \cos \rho \sin \alpha}{\sin \alpha}$$

$$= \sin \rho \, \frac{1}{\tan \alpha} - \cos \rho. \tag{3}$$

When (1) is substituted in (3), α can be eliminated to give

$$a = \frac{a}{s} \sin \rho - \cos \rho, \tag{4}$$

which provides the following central distance of *p*.

$$s = \frac{\sin \rho}{1 + (1/a) \cos \rho}. \tag{5}$$

This is a general relation for perspective projections which can be made on the equatorial plane. When $a = \infty$, (5) degenerates to

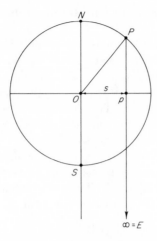

Fig. 4.   Central distance s in the orthographic projection.

$$s = \sin \rho, \tag{6}$$

the central distance for the orthographic projection (Fig. 4). When $a = 1$, (5) degenerates to

$$s = \frac{\sin \rho}{1 + \cos \rho}$$

$$= \tan \tfrac{1}{2}\rho, \tag{7}$$

the central distance for the stereographic projection, which is also derivable from the diagram of the stereographic projection in Fig. 5.

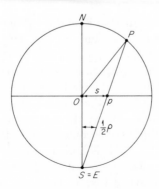

Fig. 5.   Central distance s in the stereographic projection.

Basic properties of the stereographic projection

Projections of angles. An important property of the stereographic projection is that it is angle true, that is, that any angle on the surface of the sphere is transformed into the same angle in projection. This property can be demonstrated in several ways. A simple one is the following.

In Fig. 6, suppose that any two curves on the sphere intersect at A. The curves themselves are not shown; the angle between them is defined as the angle between their tangents AT_1 and AT_2. These tangent lines are also tangent to the sphere, and so lie in the plane tangent to the sphere at A. Let these two tangents intersect the plane which is tangent to the sphere at S at the points U and V. Then, since all lines tangent to a sphere from a point are equal, the following aspects of the geometry are equal.

$$UA = US;$$

$$VA = VS;$$

$$\therefore \triangle UAV = \triangle USV;$$

$$\therefore \angle UAV = \angle USV.$$

The equatorial projection plane cuts the pyramid $USAV$ in triangle uav. Since the projection plane is parallel to the plane tangent to the sphere at S, it follows that

$$\triangle uav \text{ is similar to } \triangle USV,$$

$$\angle uav = \angle USV = \angle UAV.$$

Thus any angle (such as UAV) on the sphere projects as an equal angle (in this case uav) on the stereographic projection.

Projections of circles. Another important property of the stereographic projection is that circles on the sphere project as circles. There are several ways of demonstrating this.

In Fig. 7 there is shown a small circle on the sphere and its stereographic projection. The rays from S to the circle are the generators of an oblique cone whose base is the circle on the sphere. The stereographic projection of the circle is the conic section at the intersection of this cone and the stereographic plane. Of the possible conic sections, the parabola corresponds to a section of a cone whose axis is horizontal, and a hyperbola to a horizontal section of a cone whose axis slopes downward from S: neither of these situations corresponds to Fig. 7. Thus the stereographic projection of the base of the cone is some kind of an ellipse, possibly a circle.

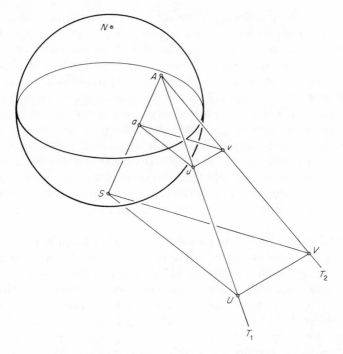

Fig. 6.   Construction used in proving that the stereographic projection is angle true.

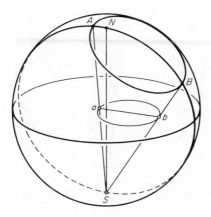

Fig. 7.   The stereographic projection of a circle.

That it must be a circle can be demonstrated elegantly with the aid of the angle-true property of the stereographic projection. Figure 8 shows the cone tangent to the sphere at the circle. The generators of this cone make right angles with the circumference of the circle. Let a plane be passed through each generator and S. These planes intersect each other in a line from the cone apex T to S, a line that passes through the stereographic plane at t. Due to the angle-true property of the projection, the right angle between every generator of the tangent cone and the circumference of the circle projects as a right angle in the stereographic projection. All the normals to the projected curve must meet at a common point t in the projection. Only that degenerate form of an ellipse which is a circle has the property that all the normals to the circumference meet in a point. Thus the circle on the sphere transforms to a circle in stereographic projection.

A meridian section [†] through these two circles which cuts them symmetrically is shown in Fig. 9. There it is seen that the projection of the center of the circle on the sphere and the center of the projected circle are not the same. Since the center of the circle on the sphere lies within the sphere, it projects to a point nearer the center of the projection. The center of the projected circle lies at the point c, but the line from the apex T, of the tangent cone, to S intersects the stereographic-projection plane at t.

Another demonstration of the projection of a circle makes use of a basic metric property of rays in the stereographic projection. Figure 10 shows a

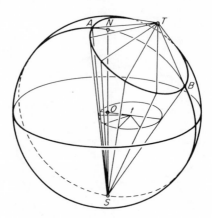

Fig. 8. Construction used in the first proof that the stereographic projection of a circle is a circle.

[†] A meridian section is a section of the sphere containing the north–south diameter.

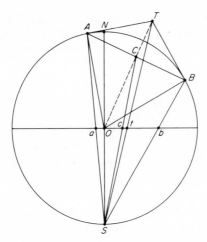

Fig. 9. Relation between the stereographic projection c of the center of a circle and the center t of the projected circle.

central vertical section of the reference sphere which contains an arbitrary point P on the surface of the sphere, and the viewing point S. The lengths of the rays from the viewing point to the point on the sphere, and from the viewing point to the stereographic projection, have interesting relations which can be derived from Fig. 10, where the stereographic projection of P is p. In the triangles SPN and SOp, the angles SPN and SOp are right

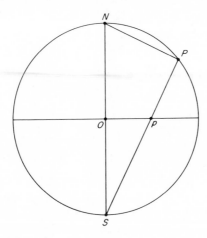

Fig. 10. Construction for demonstrating the basic metric properties of rays in the stereographic projection.

angles, and the triangles have the common angle α. Accordingly, the triangles are similar, and in them the following proportion holds:

$$\frac{Sp}{SN} = \frac{SO}{SP}. \tag{8}$$

This leads to a basic relation

$$\overline{SP} \cdot \overline{Sp} = \overline{SO} \cdot \overline{SN} = R \cdot 2R = 2R^2 = \text{constant}. \tag{9}$$

Thus, the product of lengths of the rays from the viewing point to the point on the sphere and to its stereographic projection is a constant, equal to $2R^2$, where R is the radius of the reference sphere. This is a general property of all such rays in the stereographic projection. It can be applied in evaluating the projection of a circle in the following way.

In Fig. 7 the rays from S to the circle AB determine an oblique cone with a circular base. This cone has mirror symmetry in the meridian plane $ANBS$. Let this cone be rotated 180° about its axis, so that in the meridian section (Fig. 11) $A \rightarrow A'$ and $B \rightarrow B'$. As a result of (9) the relations of the rays along the unrotated cone are

$$\overline{SA} \cdot \overline{Sa} = \overline{SB} \cdot \overline{Sb} = \text{constant}, \tag{10}$$

so that

$$\frac{SA}{SB} = \frac{Sb}{Sa}. \tag{11}$$

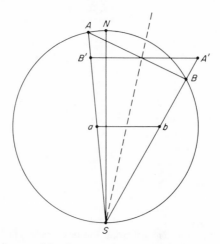

Fig. 11. Construction used in the second proof that the stereographic projection of a circle is a circle.

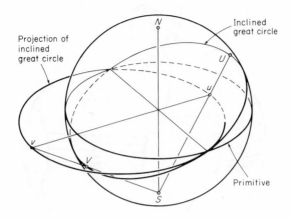

Fig. 12.

Therefore the triangles SAB and Sba are similar. Accordingly, when the cone is rotated 180°, $SA'B'$ and Sba are similar, and $A'B'$ is parallel to ba, that is, these are both horizontal.

In fact, the two cones which have the common apex S (one cone with base $A'B'$ and the other with base ba) are similar, and are so oriented that they have parallel bases. Since base $A'B'$ is a circle, so is the base ba, which is the stereographic projection of the original circle AB.

The intersection of the equatorial plane of projection with the sphere is called the *fundamental circle* or *primitive circle* (or sometimes just *the primitive*). A great circle intersects the fundamental circle in a horizontal diameter (Fig. 12). Accordingly, the projection of any great circle intersects the fundamental circle at the ends of a diameter of the fundamental circle. Unless the great circle is the fundamental circle itself, it has a projected radius greater than that of the fundamental circle, and extends beyond the limits of the fundamental circle.

Some features of crystallographic problems

Crystallographic problems involving angular relations among planes and lines can be solved quickly, and to the limit of accuracy of a graphical method, by plotting the known relations on the stereographic projection and then manipulating the projection in relatively simple ways. The entire solution can be handled in a very fundamental way by using as tools only a pencil, straightedge, compass, and protractor. Alternatively, a coordinate

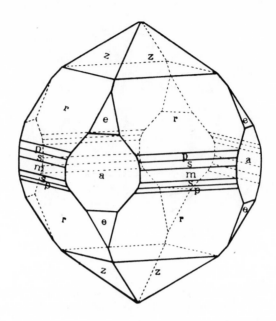

Fig. 13.   Development of faces in a crystal of anatase, TiO₂ [from A. E. H. Tutton, Crystallography and practical crystal measurement, Vol. 1 (Macmillan, New York, 1922), p. 2.]

net, known as a *Wulff net,* may be employed to achieve even more rapid solutions. The basic features of both of these methods are outlined later.

Most crystallographic problems involve crystallographic planes and crystallographic directions, occasionally even irrational planes and directions. In all cases the planes and directions are regarded as containing the center of the sphere.

With this convention, a plane can be represented on the sphere by the great circle where it intersects the sphere. It can also be represented by one of the two points where its normal intersects the sphere.

A direction may be represented by one of the two points where the line from the center intersects the sphere. It can also be represented by the plane normal to the direction; this is a great circle on the sphere. Thus a plane or a direction can be represented in two ways which bear a reciprocal relation to one another.

When the study of crystals is concerned with crystal faces, the standard practice is to represent a plane (often a crystal face) by a point where the normal intersects the surface of the sphere. This is called the *pole* of the plane. In terms of rational lattice features, every rational direction is common to an unlimited number of rational planes. The normals to these planes are also normal to the common direction, and lie in a plane normal to the common direction. In the surface development of a crystal, planes parallel to a common direction intersect one another in lines parallel to that direction, and tend to form a belt or band (sometimes interrupted) about the crystal. Such a set of faces is said to constitute a *zone*. The rational direction is called the *zone axis*. The plane containing all the normals to the faces in a zone intersects the sphere in a series of points which map out a great circle, called a *zone circle*. Thus the zone circle is the locus of the poles of the faces in the zone. Since every rational plane contains an unlimited number of rational directions, each pole is at the intersection of many possible zone circles. An example of these features, as represented in the stereographic projection, is seen in Figs. 13 and 14.

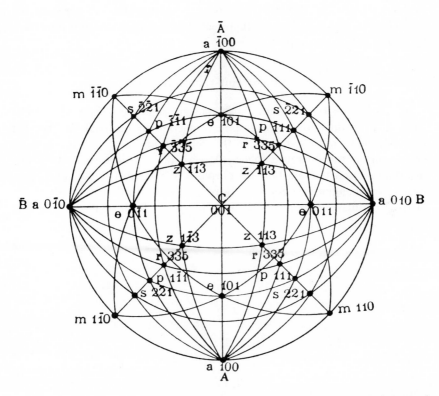

Fig. 14. Stereographic projection of the anatase crystal of Fig. 13. Each lettered face in Fig. 13 is represented in this projection by a correspondingly lettered point. Sets of faces in Fig. 13 which interesect in lines which are parallel appear in this projection as points in a common zone circle (which may appear in projection as a straight line through the center). [From A. E. H. Tutton, Crystallography and practical crystal measurement, Vol. 1 (Macmillan, New York, 1922), p. 209.]

When the angular relations between faces on a crystal were established by measurements made with a one-circle goniometer, the crystal had to be mounted so that it could be rotated about one zone axis at a time. The result of this one set of measurements was only the relative locations of the poles lying in a single zone circle. It was natural, therefore, that the crystallographic practice which grew up in the study of crystal faces should represent a crystal face as a point (pole) and an important crystallographic direction as a circle (zone). Nevertheless, there are many problems which are readily solved by the alternative convention, namely, that of representing a plane as a circle and a direction as a point on the sphere. Although the stereographic projection is independent of which convention is used to represent planes and directions, crystallographers prefer to reserve the designation "stereographic projection" for use with the representation of planes by poles and directions by zones. They designate a stereographic projection in which planes are represented by great circles and directions by points as a *"cyclographic projection"*.

Basic constructions

In this section the basic graphical constructions useful in solving problems with the aid of the stereographic projection are outlined. The general theme for many of these schemes is that one first imagines that he is looking at the top of the reference sphere. After making a decision as to which meridian section is important for purposes of the construction to be performed, one imagines the sphere to be rotated about the horizontal diameter of that meridian section, so that one can appear to draw directly in that

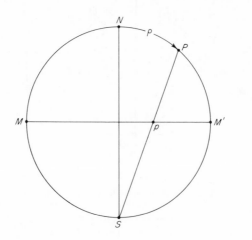

Fig. 15

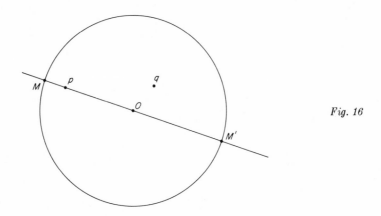

Fig. 16

meridian section. After the construction, the sphere is imagined rotated
back again about the same diameter, thus restoring the projection to its
original orientation.

  a. To plot a pole knowing its polar angle.   This is a fundamental construction. The polar angle ρ which the point P makes with the north pole
N is given. In Fig. 15, the sphere is imagined rotated 90° about MM',
the diameter of the meridian circle on which P lies. The arc ρ is laid off
from N and then projection line PS drawn. Where this line intersects
MM' is the stereographic projection p of P. The sphere is then imagined
returned to its original orientation. (More simply, the angle $\frac{1}{2}\rho = NSP$
is laid off at S in order to locate p.)

  b. To find a zone circle for any two poles.   A zone circle is a great circle,
so it must intersect the fundamental circle at the ends of a diameter of the
fundamental circle. Three points are necessary to define a circle, but
unfortunately in this case only two are given. For a third point, however,
the point on the sphere opposite one of the two given points may be used.
Given p and q (Fig. 16), the point opposite p (say), called $\bar{p}$, is required.
To find this, rotate the sphere 90° about the diameter MM' of the meridian
circle containing p (Fig. 17). Locate P and its opposite $\bar{P}$, and project $\bar{P}$
to the extended diameter at $\bar{p}$. (Or simply lay off $\angle pS\bar{p} = 90°$.) Rotate
the sphere back to its original position (Fig. 18) and draw a circle[†] through

---

  [†] Often such zone circles have such large radii that they cannot be constructed with
an ordinary compass. Circles with radii up to, say, a meter can be drawn with a beam
compass. Circles with radii up to ∞ can be drawn with a device which is essentially a
bent elastic strip whose curvature can be adjusted. This method is described in a later
section.

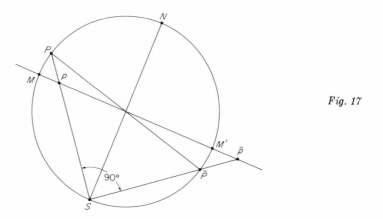

Fig. 17

the three points p, q, and $\bar{p}$. If the solution is correct, this large-radius circle will intersect the fundamental circle at the ends of a diameter of the fundamental circle.

    c. To find the pole of a zone circle. If the zone circle is known (Fig. 19), its pole (the direction of the zone axis) lies along the symmetrical meridian MM' and 90° away from the zone circle. To find this point, rotate the sphere 90° about MM', when it appears as in Fig. 20. Draw Sz, extending it to Z. Lay off $\angle ZOP = 90°$ at O, thus determining p (or more simply, lay off $\angle zSp = 45°$ at S), and then rotate the sphere back to its original position (Fig. 19).

    d. To find the equator of a pole. This is the reverse of the previous construction. In Fig. 19, pole p is given. The sphere is imagined rotated 90°

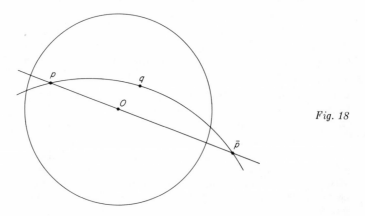

Fig. 18

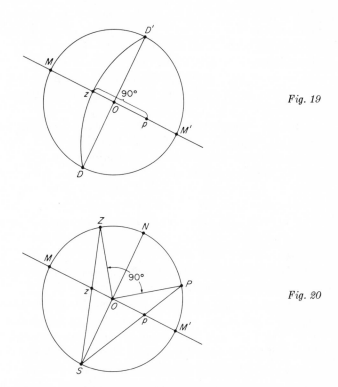

Fig. 19

Fig. 20

about the diameter MM' to give Fig. 20. Draw Sp, extending it to P. A right angle POZ is laid off at O (or, more simply, $\sphericalangle pSz = 45°$ at S). After the sphere is rotated back to the condition of Fig. 19, the point z is known. The zone circle must pass through z and also intersect the fundamental circle at D and D'; through these three points an arc can be readily drawn.

e. To find the angle between any two poles. The background for this construction can be understood from Fig. 21. The magnitude of the arc PQ is required from the location of the points p and q on the stereographic projection. These points on the sphere are members of a zone circle $DPQD'$, whose normal OZ is defined by $\sphericalangle NOZ = \psi$ at the center of the sphere. Pass a great circle halfway between zone circle $DPQD'$ and the fundamental circle, so that the three great circles intersect in diameter DD' and have angular separation $\frac{1}{2}\psi$. The normal to the halving plane is SZ, which makes angle $\frac{1}{2}\psi$ with SN. Pass two planes through line SZ, one containing P, the other Q. These planes intersect the sphere in small circles ZPS and ZQS. Their intersections with the fundamental circle are the two lines

zP' and zQ'. Now, the halving plane relates the features in the zone circle and fundamental circle as if the halving plane were a mirror. The arc PQ in the zone circle is wanted. This is the same as $P'Q'$ in the fundamental circle. Points P' and Q' are at the intersections with the fundamental circle of lines zp and zq (where z is the stereographic projection of the pole of the zone circle, and p and q are the stereographic projections of the poles in question).

This theory leads to the following construction (Fig. 22). Given p and q in the stereographic projection. To find the magnitude of the arc between them, first find the zone circle connecting them as in construction b, and its pole z as in construction c. Draw zp and zq, extending them until they reach the fundamental circle at P' and Q'. The intercept arc is equal to the great-circle arc between the poles P and Q, of which p and q are the stereographic projection. The point z is sometimes called the *angle point* of the zone circle pq.

f. To locate a pole in a zone circle, given its angle to another pole in the zone. This is a corollary to construction e. The zone circle, and some pole along it, say p (Fig. 22), are given. First find the zone pole z. Draw zp, continuing it to P'. Lay off arc $P'Q'$ along the fundamental circle equal to the angle between the two poles, and in the correct direction. Draw zQ'. The intersection q of this line with the zone circle is the required pole.

g. To find the angle between two zone circles. The obvious way to find the angle between two zones is to draw tangents to the two circles at their intersection and then, making use of the angle-true property of the stereographic projection, measure the angle between these tangents. The difficulty with this straightforward method is that it is not easy to draw tangents precisely.

An alternative and more precise construction is to regard the point of intersection of the two zone circles as a pole a (Fig. 23); this can be precisely located. Let the equator of pole a be found by the method of construction d, illustrated in Figs. 19 and 20. This establishes, in Fig. 23, great circle EE' which is everywhere 90° from pole a; therefore the true magnitude of arc bc on the sphere measures the angle between the two zone circles. The actual size of this arc is $b'c'$, according to the construction of a foregoing section.

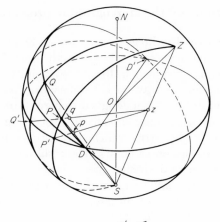

Fig. 21

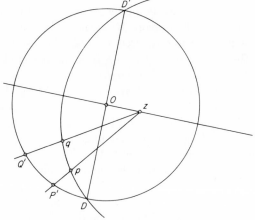

Fig. 22

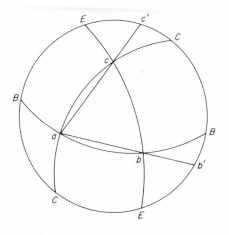

Fig. 23

Aids to using the stereographic projection

Instruments for drawing large-radius arcs. Zone circles in the stereographic projection have radii whose lengths vary from that of the primitive circle (commonly about 5–10 cm) to infinity. Those with the larger radii are impossible to draw with ordinary drawing instruments, so several auxiliary devices have been designed to permit drawing them. Two simple devices are shown diagrammatically in Figs. 24 and 25.

Figure 24 shows a flexible strip, which may be metal or plastic, bent between two inner and two outer points. The resulting curve of the strip is a good approximation to a circular arc of large radius. The radius is adjustable by shifting the plate carrying the inner points. The two plates can be locked together for drawing purposes. This device was first proposed by Wulff[5] and improved by Fedorov.[6]

Large-radius circles can also be perfectly sketched by making use of an elementary theorem of plane geometry: An angle whose apex is on the circumference of a circle is measured by half the intercepted arc. The application of this theorem to drawing large-radius circles is illustrated diagrammatically in Fig. 25i. The circle is the zone circle; the angle in question is μ, and the intercepted arc is AWB. Using the theorem in a reverse way, let A and B be fixed points whose linear distance apart AB is equal to the diameter of the primitive circle of the stereographic projection. If a triangle MNQ, whose apex angle at N is μ, is moved so that side MN maintains contact with point A and side NQ maintains contact with point B, then point N accurately moves along the arc AB of a very large zone circle. The mechanical device for realizing this may be as simple as the three sticks shown in Fig. 25ii. This principle was introduced to crystallography by Goodchild[7], Staring[21], and Herlinger[22].

Scales and protractors. The most important tool for many problems involving the stereographic projection is a scale showing the linear distance

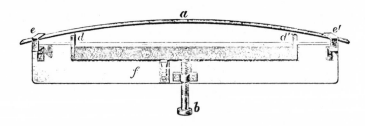

Fig. 24. Ruler for drawing circular curves of large radius. [From Wulff[5], pp. 253–254.]

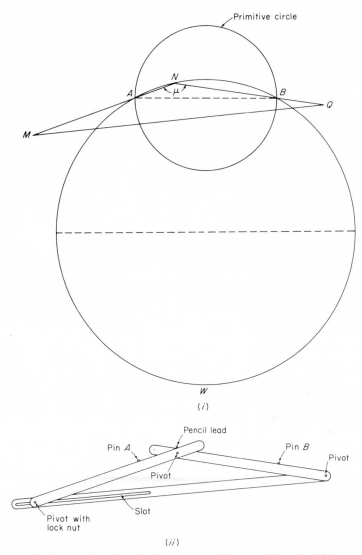

Fig. 25. Device for drawing circular curves of large radius. (*i*) Diagram of the geometry. (*ii*) A simple device for applying (*i*).

of a pole p from the north pole of the projection as a function of the angle ρ between the pole and the north pole. This is simply a graphical solution of (7) but enlarged by a factor equal to the radius of the fundamental circle in whatever linear units are used. This is commonly 5 or 10 cm, but may be any convenient radius expressed in any convenient unit of length.

(i)

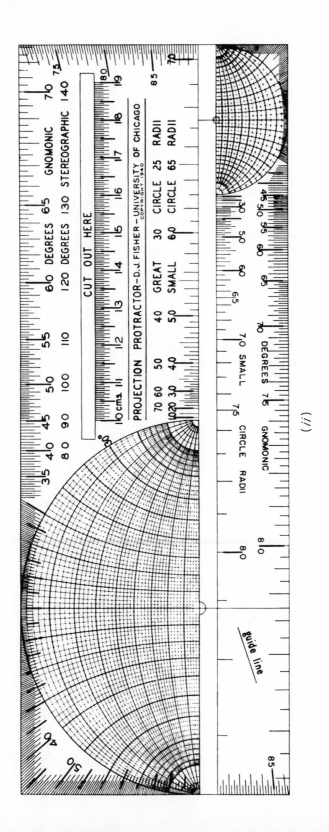

Fig. 26. "Protractors" for aiding stereographic sketching. (*i*) From T. V. Barker[20], p. 1; (*ii*) from D. Jerome Fisher[29], p. 295.]

Such a scale is usually combined with other useful information, such as the distance to the gnomonic-projection point for the same pole, and with miscellaneous data, such as radii of small circles and half a Wulff net (described in the next section). Such combination devices are called *protractors*. T. W. Barker's protractor is shown in Fig. 26*i* and D. J. Fisher's version in Fig. 26*ii*.

The Wulff net and its use. In solving problems with the aid of the stereographic projection, many of the geometrical constructions can be avoided if a coordinate system is used. A standard way of referring to points on a sphere is by means of latitude and longitude. The parallels of latitude are, except for the equator, small circles, whereas longitude is marked out by great circles. Using the standard constructions of the stereographic projection, it is an easy but tedious matter to map the latitude and longitude circles in stereographic projection. Two such maps of nets are shown in Figs. 27 and 28. The first, known as the *polar stereographic net*, is the net of latitude and longitude lines of one hemisphere, as stereographically projected from the south pole; the second, as projected from a point on the equator.

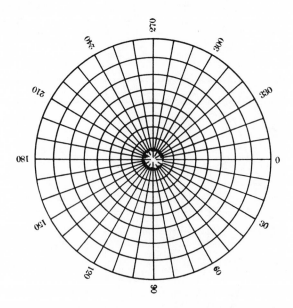

Fig. 27. Polar net of latitude and longitude lines for the stereographic projection as projected on the equatorial plane (from the south pole as viewing point). [From T. V. Barker[20], p. 14.]

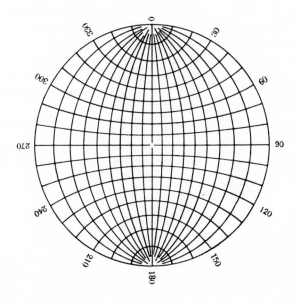

Fig. 28. Equatorial (Wulff) net of latitude and longitude lines for the stereographic projection as projected on a meridian plane (from a viewing point at 90° to this plane). [From T. V. Barker[20], p. 15.]

The net shown in Fig. 28, known as the *Wulff net* or the *equatorial stereographic net*, is a standard tool for the solution of problems involving the stereographic projections. Some of the useful properties of the stereographic projection are brought out by it. All the latitude and longitude lines are circles, which is a consequence of the fact that all circles, great and small, appear on the stereographic projection as circles. Furthermore, all latitude and longitude lines on the sphere cross each other at right angles; on the Wulff net their projections cross at right angles also. This is a consequence of the angle-true property of the stereographic projection. The general adjective for a map that preserves angles is *conformal*. Furthermore, in any tiny region of the stereographic projection, a square projects as a square, a circle as a circle, etc., so the stereographic projection is said to be orthomorphic, that is, it preserves shapes locally, which is a useful attribute.

On the other hand, some disadvantages are also obvious from the Wulff net: a small segment of arc on the sphere appears in stereographic projection as its actual length when at the fundamental circle, but as only half its true length at the center of the projection. Small areas are therefore correctly represented on the fundamental circle, but are projected as only a quarter of their true area at the center.

Some features of the Wulff net are provided by the stereographic protractors. In particular, the centers and radii of small circles are given by Barker's scale B and the corresponding part of Fisher's protractor.

To use the Wulff net, a piece of tracing material is placed over it, and the center of the net carefully marked on the tracing material. There are two basic operations which can be readily performed with the aid of the net.

1. The drawing and net may be rotated with respect to each other. This operation can be carried out most easily by using the special drawing board devised by Wülfing[14] and improved by Johannsen[15]. With this tool, the net may be rotated beneath the drawing (or, in the commercial version by Leitz, the drawing may be rotated over the net).

A simple use of this rotation is to permit the angle between any two points to be measured. To accomplish this, the drawing is rotated until the two points lie on the same meridian of the net, which is a great circle. Then the angle between the two points is their difference in latitude.

In this position the meridian on which the two points lie is also their zone circle. The pole of this zone can be easily found. It lies on the equator of the net, 90° away from the point where this meridian cuts the equator.

2. The sphere, of which the net is the stereographic projection, may be imagined rotated about an axis in the plane of the net. Let this rotation

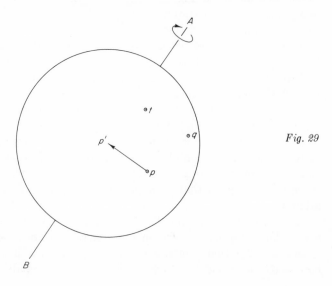

Fig. 29

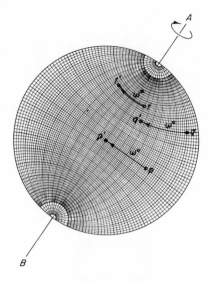

Fig. 30

axis be taken as the line connecting the north and south poles on the Wulff net. If the sphere is rotated about this north–south axis, every point on the sphere migrates along a constant-latitude line. If the sphere is rotated by $\omega°$, each point migrates (along a constant-latitude line) to a meridian line of $\omega°$ greater (or lesser) value.

For example, Fig. 29 shows a stereographic projection with three points p, q, and t. Suppose it is necessary, in some problem, to imagine the sphere rotated so that p migrates to the center of the stereographic diagram. This would be accomplished by rotating the sphere about an axis AB perpendicular to the desired path pp'. But as $p \to p'$, what happens to q and t? Rotation of the sphere about AB causes all points to migrate along parallels of latitude, which are the small circles in Fig. 30. The angular amount of migration is ω, the angle necessary to bring p to p'. This angle is measured by the angle between meridians, and can be read directly on the Wulff net.

Example of the use of the stereographic projection[27]

An example of the application of the stereographic projection in solving problems of a general nature which occur in the field of crystallography is the following. Certain crystals can dissolve foreign material when heated, and can expel it again when the temperature is reduced. Sometimes the exsolved material is a plate-shaped crystal which has a definite orientation with respect to the crystallographic directions of the crystal that expels it.

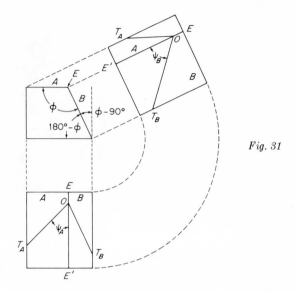

Fig. 31

To discover what this relation is, one needs some angular data that relate the orientations of the plate to the orientation of two natural or artificial faces of the host crystal. For example, the traces made by the plate as it crosses surfaces of the host crystal can be measured. The geometry involved is illustrated in Fig. 31. The two adjacent faces of the host crystal are A and B, which make angle ϕ with one another. The thin plate intersects

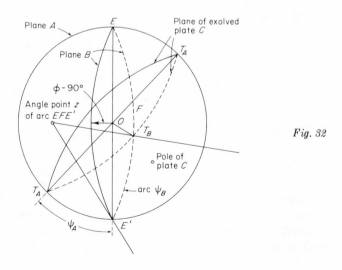

Fig. 32

faces A and B as traces OT_A and OT_B, which make angles ψ_A and ψ_B with the common edge. The question is, what is the orientation of the plane of the plate with respect to A, B, and the edge EE between them?

The stereographic solution is given in Fig. 32. In this projection the planes and lines meeting at O in Fig. 30 should be imagined as translated to O in the sphere and stereographic projection. In cyclographic representation, plane A is taken as the fundamental great circle, and the edge between A and B is points E and E' in this plane. The plane of B is the great circle whose diameter is EE' and which makes an angle ϕ with the fundamental circle. Line OT_A is laid off, making ψ_A with EE. Line OT_B lies in plane B and makes an angle ψ_B with OE'. The required plane C, therefore, is a great circle containing lines T_A and T_B. Accordingly, a great circle is drawn through T_A and T_B. This is the representation of the required plane C of the exsolved plate.

Crystal drawing

It is customary to represent the appearance of a crystal by an orthographic projection, that is, by a drawing of the way the crystal would appear from a point sufficiently distant so that there is no obvious perspective. In this way, parallel lines on the crystal appear parallel in the drawing of the crystal. The crystal drawing itself consists essentially of a number of straight lines, each line indicating the intersection of two neighboring faces of the crystal. A basic feature of a crystal drawing, therefore, is the direction of each of these lines, which are zone axes. This information can be derived directly from the stereographic projection [†] of the crystal.

The basis for this is shown in Fig. 33. The locations a and b, in the stereographic projection, of two faces on the crystal are given, and the direction of their intersection on the drawing of the crystal is desired. These faces lie in a zone whose stereographic projection is the great circle $DabD'$ and whose pole is p. If the sphere is rotated 90° about Op, the north pole migrates to N and the south pole to S. Then the zone circle on the sphere appears as the straight line OZ whose normal is the zone axis OP. The stereographic projection of the zone axis OP is Op, and its orthographic projection is Op'. The lines OP, Op, and Op', accordingly, lie in the same meridian plane and project along the line OM, which in the stereographic projection is normal to the chord DD' of the zone circle $DabD'$. As seen by

---

[†] In the next chapter it will be seen that the information is also available from the gnomonic projection.

the eye looking at the stereographic projection from an infinite distance from the side of the north pole, the edge between faces whose projections are poles a and b in the stereographic projection is simply the normal OM to the chord of their zone circle.

On an actual stereographic projection, zone circles can be drawn through *any* two faces. But the only zone circles that are needed for the crystal drawing are those which connect faces that are seen to intersect in a line on the actual crystal. Ordinarily this is a small number. A drawing of a crystal made using this method is shown in the upper right of Fig. 34. Such a drawing is always the way the crystal, oriented as mapped in the stereographic projection, appears when the eye looks along the line NS

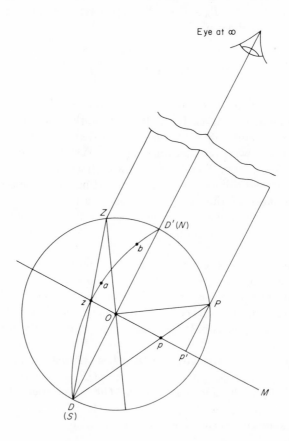

Fig. 33.   The basis for using the stereographic projection for making drawings of crystals.

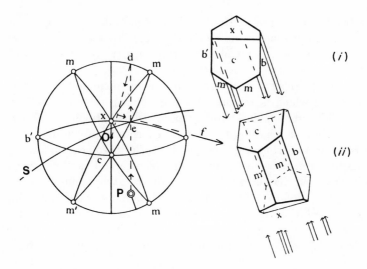

Fig. 34i. Example of the use of the stereographic projection in crystal drawing: orthoclase, seen looking along the *c* axis.

Fig. 34ii. Drawing of the crystal in Fig. 34*i* from a more general viewpoint. [From Barker[20], p. 52.]

of the sphere from a considerable distance. (The Germans call this a *Kopfprojektion*, or projection looking at the "head", or end, of the crystal.)

A drawing can be made looking at the crystal from any direction by simply rotating the sphere, as well as the face-pole locations and viewing point on its surface, until the viewing point occupies the center of the stereographic projection. This can be readily accomplished with the aid of the Wulff net[32,33].

A crystal drawing from an arbitrary viewing point can readily be made without using the Wulff net. The principle is illustrated[†] in Fig. 35*i*. The viewing direction is indicated by the point *w*, whose angular coordinate

[†] In Fig. 35*i*, full lines are the stereographic projections of elements on the upper hemisphere as projected from the south pole, while dashed lines are the stereographic projections of elements on the lower hemisphere as projected from the north pole. This dual-projection scheme, commonly used in stereographic projections, has the advantage of limiting the area of the projection of the entire sphere to the fundamental circle. The result resembles the orthographic projection of the whole sphere, in which a great circle would appear as an ellipse. In this kind of stereographic projection this full circle actually appears as two circular arcs (one projected from each pole) with a common diameter. Thus *AdBd'* in Fig. 35*i* consists of the full arc *AdB* and the dashed arc *Bd'A*, which together resemble an ellipse. This would be a true ellipse on the orthographic projection of the sphere.

from a front view in the horizontal plane is ϕ, and whose angular coordinate from the horizontal (the equator) is ψ. This requires the crystal to be projected on the equator of this direction, namely, on the great circle AdB.

Let the great circle UV be the zone circle corresponding to the interfacial-edge direction which is required for the viewing direction indicated by point w. This problem has already been solved provided the sphere is rotated so that w migrates to O, that is, rotated about diameter AB. When rotation is performed, great circle $AdBd'$ becomes the fundamental circle. Before rotation, the arbitrary zone circle UV and the projection plane intersect in diameter dOd'. After the rotation, zone circle UV migrates to great circle $EF'E'$, diameter dOd' becomes EOE', and all points on great circle UV on the side toward w come to occupy the upper hemisphere, that is, the arc EE' outlined in a solid line, while those points on great circle UV on the side away from w come to occupy the lower hemisphere.

Since w is the pole of great circle AdB, it is the angle point for the circle, according to construction e. Therefore, if line wd is drawn and extended to E, arc $AE =$ arc Ad. Thus EOE' is the required diameter of the zone circle in the rotated position, and the normal to it, namely tangent TE, is

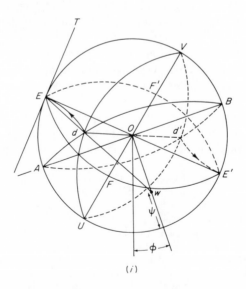

(*i*)

Fig. 35i. The geometry used in demonstrating how the direction of a zone line in a crystal drawing may be determined from the stereographic projection.

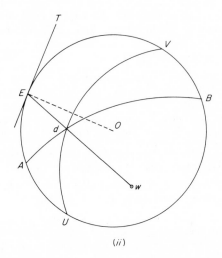

Fig. 35ii. The actual construction needed to derive a zone line in a crystal drawing from the stereographic projection.

the required direction of the edge between any two intersecting faces in the zone *UV*.

This discussion is concerned with the demonstration of the correctness of the construction. The routine construction required is very limited, and illustrated in Fig. 35*ii*. The viewing direction *w* and the projection plane *AB* need be constructed but once for the entire drawing. Then a zone circle, such as *UV*, must be drawn for those faces that actually intersect. For each such circle, its intersection *d* with the projection circle must be located; then line *wdE* drawn; then the tangent *TE* at *E* drawn.

A standard viewing direction usually assumed for crystal drawing is one that is to the right of the prime meridian (Fig. 34) by $\phi = 18°26' = \tan^{-1}\frac{1}{3}$, and above the equator by $\psi = 9°28' = \tan^{-1}\frac{1}{6}$. These values are heritages of an earlier practice of drawing crystals by the method known as the *clinographic projection*, which called for first drawing a coordinate system (an "axial cross") approximately as seen from a convenient viewing direction, and then fitting the crystal planes to this set of axial lines so as to intercept them at appropriate distances.

The methods outlined above, however, are suitable to any viewing direction. Often a particular viewing direction gives rise to an awkward drawing. For example, in Fig. 35, suppose the pole of a face coincides with intersection *d*. Then, in rotated position, the pole is *E* and the face itself appears as straight line *TE* instead of an area. A face appears almost as a straight line if it is on the zone circle near d. In such instances a better-appearing drawing can be made by choosing a different viewing direction.

Summary

Crystallographers occasionally carry out certain calculations involving angles among planes and directions with the aid of special projections. As a preliminary, the planes and directions are moved to the center of a sphere, called the *reference sphere*. The intersections of the planes and directions (or their perpendicular lines and planes, respectively) with the surface of the reference sphere constitute the *spherical projection* of these geometrical elements.

The surface of the reference sphere and the distribution of markings on this surface, as seen from various eye positions, constitute a *perspective projection* of these items. If the eyepoint is at an infinite distance from the sphere, the result is the familiar *orthographic projection*, whereas if the eyepoint is at the south pole, the result is the *stereographic projection*. In this projection, a point whose angular position from the north pole is ρ appears at a distance of $R \tan \frac{1}{2}\rho$ from the center of the projection (where R is the radius of the reference sphere).

The stereographic projection has the basic property that any angle between two lines on the sphere projects as the same angle, so that the projection is angle true (conformal). Another useful property, which can be derived from the angle-true characteristic, is that circles on the sphere project as circles.

When the surface development of a crystal is studied, a crystal face is represented on the spherical projection by the point (called the *pole* of the face) where the face normal intersects the sphere's surface. A set of faces that are parallel to a common direction is called a *zone*, and their poles lie on a great circle, the *zone circle*, which is normal to the common direction, which is called the *zone axis*. These poles, zone circles, and zone axes are readily plotted in the stereographic projection by one of two means: they may be plotted, by using pencil, straightedge, compass, and protractor, from data on the angles between poles in zones, or they may be plotted from the angular coordinates of the faces by using a *Wulff net*. Problems involving angles among planes and directions can be solved by using either method.

Crystal drawings can be readily made by deriving the directions of the edges between adjacent faces in a zone from the stereographic projection. The crystal can be drawn as seen from any chosen viewing direction.

Notes on history

The stereographic projection, as well as its use in representing the surface of a sphere, was known as early as the second century B.C., when it was probably devised by the Greek astronomer Hipparchus. In the second century after Christ, the famous Alexandrian astronomer and geographer, Claudius Ptolemy, who presumably learned of it from Hipparchus, described it in his book. This work was later translated into Latin by Fredericus Commandinus, who published it in Venice in 1558 under the title *Ptolemaei planisphaerium.* There is no certainty who discovered and proved the angle-true nature of the stereographic projection; it appeared to have been common knowledge as early as 1700, and may have been first demonstrated by Edmund Halley in 1695. The technique of projecting circles as circles was probably known already to Ptolemy.

The early uses of the projection were in making geographic and astronomic maps and for solving spherical triangles. The stereographic projection is still used to a limited extent in making geographic maps.

This projection was introduced into crystallography by Neumann in his book *Beiträge zur Kristallonomie* in 1823. In his Fig. 20 he reproduced his stereographic projection of *Rothgiltigerz* (pyrargyrite) and in Fig. 29 a corresponding projection of vesuvianite. With the publication of Miller's *Treatise on crystallography,* and later of Des Cloizeaux's *Manuel de mineralogie,* its application in crystallography and mineralogy was popularized, and it has been used for representing all kinds of angular relations in crystals, especially minerals, ever since. It was most extensively used in the days preceding the first world war to represent the locations of poles of crystal faces.

Because of its angle-true property, the stereographic projection has been frequently applied to the solution of problems involving angular relations of interest in crystallography, and it is still used to some extent for this purpose. The chief virtue of the stereographic projection in this connection is that it permits quick solutions of such problems with the accuracy permitted by a graphical method. Graphical solutions were rendered easy by Wulff's introduction, in 1902, of a net [†] that made it possible to plot, on the stereographic projection, points whose angular coordinates were known on the reference sphere. This application became even easier with the aid of a special drawing table devised by Johannsen, and later manufactured by the Leitz optical company, which permitted easy rotation of the drawing paper about the center of the Wulff net. The graphical solution

---

[†] This had been anticipated by Adranus Metius in his *Primum mobile,* published in Amsterdam in 1633.

of problems with the stereographic projection has declined since good digital desk computers, and, more recently, high-speed electronic computers, have become available.

The stereographic projection is still commonly employed to illustrate simple angular relationships in crystallographic publications, for example, to display the space relations between axes and planes of symmetry. This seems to be due to the influence of the crystallographic computing practice of an earlier era. Such use of the stereographic projection for illustrations, in which nothing is to be measured, is undesirable because the eyepoint of this perspective projection is a most unusual one which produces a picture of unnatural appearance.

Significant literature

1. F. E. Neumann. Beiträge zur Krystallonomie. (Ernst Siegfried Mittler, Berlin and Posen, 1823) 152 pages and 43 Figs. (12 Tafeln "in Steindruck").
   [Repeated by C. Neumann: *Abh. K. Sächs. Ges. Wiss. Math.-Phys.* (1917) 195–458.]
2. W. H. Miller. A treatise on crystallography. (J. and J. J. Deighton, Cambridge, 1839) 139 pages and 10 plates, especially 129–139.
3. A. Des Cloizeaux. Manuel de Minéralogie. (Dunod, Paris)
   Tome premier (1862) 572 pages with 22 stereographic projections of minerals, and an atlas of 52 plates with 313 figures.
   Tome second, 1er fascicule (1874) LIV plus 208 pages with 7 stereographic projections of minerals and an atlas of 16 plates and 101 figures.
   2ème fascicule (1893) pages 209–544, with 11 stereographic projections of minerals and an atlas of 16 plates and 100 figures.
4. Victor Goldschmidt. Ueber Projection und graphische Krystallberechnung. (Julius Springer, Berlin, 1887) 97 pages, especially 36–53.
5. Georg Wulff. Ueber die Vertauschung der Ebene der stereographischen Projection und deren Anwendungen. *Z. Kristallogr.* **21** (1893) 249–254, especially 253–254.
6. E. von Fedorow. Universal- (Theodolith-) Methode in der Mineralogie und Petrographie. *Z. Kristallogr.* **21** (1893) 574–678.
7. J. G. Goodchild. Simpler methods in crystallography, Part I. Stereograms. *Proc. Roy. Phys. Soc. Edinburgh* **14** (1900) 323–359.
8. S. L. Penfield. The stereographic projection and its possibilities from a graphical standpoint. *Am. J. Sci.* **11** (1901) 1–24, 115–144.
9. S. L. Penfield. Ueber die Anwendung der stereographischen Projection. *Z. Kristallogr.* **35** (1901) 1–24.
10. S. L. Penfield. On the solution of problems in crystallography by means of graphical methods, based upon spherical and plane trigonometry. *Am. J. Sci.* **14** (1902) 249–284.
11. E. von Fedorow. Ueber die Anwendung des Dreispitzzirkels für krystallographische Zwecke. *Z. Kristallogr.* **37** (1903) 138–142, especially 142.
12. A. Hutchinson. On a protractor for use in constructing stereographic and gnomonic projection of the sphere. *Mineralog. Mag.* **15** (1908) 93–112, especially "Historical appendix," 105–111.

13. A. Hutchinson. Ein Transporteur für die stereographische und gnomonische Projection. *Z. Kristallogr.* **46** (1908) 225–244.
14. E. A. Wülfing. Wandtafeln für stereographische Projektion. *Centralbl. für Min., Geol. u. Pal.* (1911) 273–275.
15. Albert Johannsen. A drawing-board with revolving disk for stereographic projection. *J. Geol.* **19** (1911) 752–755.
16. H. E. Boeke. Die Andwendung der stereographische Projektion bei kristallographischen Untersuchungen. (Gebrüder Borntraeger, Berlin, 1911) 58 pages.
17. Albert Johannsen. Manual of petrographic methods. (McGraw-Hill, New York, 1918) 5–28.
18. Stimson J. Brown. Trigonometry and stereographic projections. (Lord Baltimore Press, Baltimore, 1919) 112–148.
19. Charles H. Deetz and Oscar S. Adams. Elements of map projections with applications to map and chart construction. U.S. Coast and Geodetic Survey Special Publication No. 68 (Washington, 1921) 163 pages.
20. T. V. Barker. Graphical and tabular methods in crystallography. (Thomas Murby, London, 1922) 152 pages.
21. A.-J. Starling. Un instrument pour tracer les grands cercles et mesurer des angles dans les projections stéréographiques. *Archives des sciences phys. et nat. (Genéve)* **6** (1924) 39–46.
22. E. Herlinger. Über eine Vorrichtung von Herstellung von stereographischen Netzen. *Z. Kristallogr.* **67** (1928) 543–546.
23. Robert L. Parker. Kristallzeichnen. (Gebrüder Borntraeger, Berlin, 1929) 112 pages [with 15 literature citations on crystal drawing].
24. F. E. Wright. The preparation of projection diagrams. *Amer. Mineral.* **14** (1929) 251–258.
25. Georg Scheffers. Wie findet und zeichnet mat Gradnetze von Land- und Sternkarten. (Teubner, Leipzig und Berlin, 1934) 98 pages.
26. H. Tertsch. Das Kristallzeichnen auf Grundlage der stereographischen Projektion (Julius Springer, Vienna, 1935) 38 pages.
27. Charles S. Barrett. The stereographic projection. *Trans. Am. Inst. Min. Met. Eng.* **124** (1937) 29–58.
28. F. W. Sohon. The stereographic projection (Chemical Publishing Co., Brooklyn, New York, 1941) 210 pages.
29. D. Jerome Fisher. A new projection protractor. *J. Geol.* **49** (1941) 292–323; 419–442.
30. George P. Kellaway. Map projections. (Methuen, London, 1946) 127 pages.
31. N. N. Padurow. Zweinetzmethode in stereographischer Projektion. *Neues Jahrbuch f. Mineralogie etc.* (1949) 173–176.
32. Haymo Heritsch. Über die Herstellung porträtgetreuer Kristallbilder aus der stereographischen Projektion. *Tschermaks Min. u. Petrogr. Mitt.* **2** (1950) 67–82.
33. H. Meixner. Ein "erweitertes Wulff'sches Netz" als Hilfmittel beim Kristallzeichnen. Radex-Rundschau, (1953) 51–53.
34. Hermann Tertsch. Die stereographische Projektion in der Kristallkunde (Verlag für angewandte Wissenschaften, Wiesbaden, 1954) 122 pages.
35. P. Terpstra and L. W. Codd. Crystallometry (Longmans Green, London, 1961) 420 pages, especially 11–45.
36. D. J. Dyson. A simple stereographic plotting table. *Z. Kristallogr.* **122** (1965) 307–310.
37. Charles H. Cotter. The astronomical and mathematical foundations of geography. (American Elsevier, New York, 1966) 244 pages, especially 173–233.

38. J. H. Palm. The computation of stereographic projections. *Z. Kristallogr.* **123** (1966) 388–390.
39. G. K. Stokes, S. R. Keown, and D. J. Dyson. The construction of stereographic projections by computer. *J. Appl. Cryst.* **1** (1968) 68–70.
40. R. L. Cunningham and Joyce Ng-Yelim. An optical device for the direct production and viewing of stereographic and gnomonic projections. *J. Appl. Cryst.* **1** (1968) 320–321.
41. M. J. Oppenheim. The stereographic projectionarium. *Mineralog. Mag.* **37** (1969) 524–525.
42. O. Johari and G. Thomas. The stereographic projection and its applications. (Wiley-Interscience, New York, 1969) 132 pages.
43. Marisa Canut-Amoros. STLPLT—CalComp plot of crystallographic projections of Laue photographs. *Computer Physics Communications* **1** (1970) 293–305.

Chapter 3

The gnomonic projection

Nature of the gnomonic projection

The general use of the gnomonic projection is to represent, in projection, features that occur on the surface of a sphere. It is one of the perspective projections; the eyepoint is located at the center of the sphere, and the projection plane is tangent to the sphere, the point of tangency ordinarily being taken as the north pole. The relations in the meridian plane which contains the center of the sphere O and the point P on the surface of the sphere are shown in Fig. 1. The location of point P on the sphere is defined in the meridian plane by its polar angle ρ. Its projection p occurs at a distance from the center C of the projection given by

$$r = R \tan \rho \tag{1}$$

where R is radius of the reference sphere.

The adjective *gnomonic* is derived from the Greek *gnomon*, an object like the pointer on a sundial, the length and position of whose shadow indicates the hour of the day. The gnomonic projection is geometrically similar to an upside-down sundial: the plane of the sundial is the plane of the projection; the pointer bar perpendicular to it corresponds to the line CO in Fig. 1; the ray along the top of the shadow cast by the pointer bar corresponds to the projection line OPp in Fig. 1.

A plane through the center of the reference sphere (Fig. 2) intersects the sphere in a great circle and intersects the projection plane in a straight

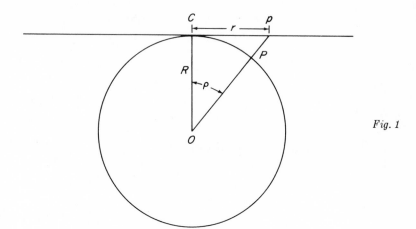

Fig. 1

line. An obvious characteristic of the gnomonic projection is, therefore, that great circles on the sphere project as straight lines. In many crystallographic uses of the projection, a point on the sphere is the pole of a crystal face and the great circle is a zone to which the face pole belongs, so in the gnomonic projection the straight line is a zone line on which the projection

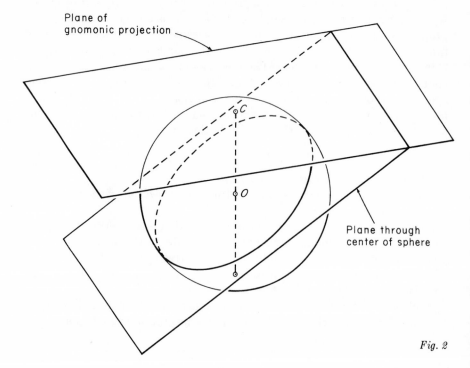

Fig. 2

of the face pole lies. The fact that zones appear as straight lines, and possible faces as their points of intersection, makes the gnomonic projection especially valuable for certain crystallographic purposes.

Whereas great circles on the sphere project as straight lines, small circles project as conic sections. This is because the projection point (the center of the sphere) and the small circle determine a circular cone which intersects the projection plane as a conic section. The intersection is a circle if the axis of the cone is vertical, an ellipse if the lowest generator points above the horizontal, a parabola if the axis is horizontal, and a hyperbola if the axis slopes downward from the center.

Indexing

The gnomonic projection came into extensive use in crystallography largely because it provided a simple graphical tool for indexing[18] crystal faces. This is possible because if a crystal axis is oriented perpendicular to the projection plane, the points representing rational planes fall on nodes of a main net, or on nodes of nets which are described by subcells of the main net. This useful characteristic is a consequence of the certain similarities in the geometries of the gnomonic projection and the reciprocal lattice.

The basic features of this relation are demonstrated in two dimensions in Fig. 3, which shows a two-dimensional lattice with one cell shaded; one of the cell axes, labeled c, is oriented perpendicular to the projection plane. One lattice point is placed at the projection center O (this is the center of the reference sphere, the existence of which is unimportant unless points on the surface of the sphere are to be projected). Several rational lattice lines OU, OV, and OW are shown which intercept the b axis at 1 unit, and the c axis at 0, 1, and 2 units, respectively. The intersections with the projection plane of the normals to these planes determine the poles u, v, and w, respectively. These poles will now be shown to be equally spaced on the projection plane, and to be the nodes of a one-dimensional lattice u, v, w, ... That the separations between the points of this sequence are equal is readily appreciated if it is noted that, for example,

$$\triangle Ouv \text{ is similar to } \triangle OUV$$

$$\therefore \frac{uv}{UV} = \frac{Ou}{OU}$$

$$\therefore uv = Ou \cdot \frac{UV}{OU} = Ou\left(\frac{c}{b}\right)$$

similarly,
$$uw = Ou \cdot \frac{UW}{OU} = Ou \cdot 2\left(\frac{c}{b}\right) \tag{2}$$

$$ux = Ou \cdot \frac{UX}{OU} = Ou \cdot 3\left(\frac{c}{b}\right)$$

$$\vdots$$

Thus, the points u, v, w, x, . . . are equally spaced, and occur at distances 0, 1, 2, 3, . . . times c/b from the point which is the projection u of the pinacoid face OU. These points correspond to lines whose intercepts are 0, 1, 2, 3, . . . on the c axis, and whose intercepts on the b axis are unity, and therefore whose indices are the intercepts interchanged, namely, (01), (11), (21), (31), ...

If, in Fig. 3, instead of a line whose intercept on c is an integer, a more general line such as OS is considered, it is seen that the following corresponding relations hold.

$$\triangle Ous \text{ is similar to } \triangle OUS \text{ and to } \triangle OU'S'$$

$$\therefore \frac{us}{U'S'} = \frac{Ou}{OU'}$$

$$\therefore \quad us = Ou \frac{U'S'}{OU'} = Ou \frac{2c}{3b} = Ou \frac{2}{3}\left(\frac{c}{b}\right)$$

The line has intercepts $3(b)$, $2(c)$, corresponding to indices (23). If referred to a pair of indices such that the last is normalized to unity, this corresponds to $(\frac{2}{3}1)$.

For two dimensions this discussion demonstrates the following useful characteristic of the gnomonic projection. When the c axis of a lattice is oriented normal to the projection plane, and when a pole u in Fig. 3 is chosen for the basal pinacoid and the pole v is chosen as the unit pyramid, then the distance of any pole from the pole u of the basal pinacoid gives the first index of the pole relative to a second index taken as unity.

The generalization to three dimensions is illustrated in Fig. 4 by a view looking along line CO of Fig. 3. Each vertical line of lattice points in Fig. 3 is perpendicular to the projection plane, and one line would intersect it at C of Fig. 4. The plane of Fig. 3 is the horizontal line $Cuvwx$ of Fig. 4. The locations of poles u, v, w, . . . is still valid when the a axis of the lattice is perpendicular to the plane of axes b and c of Fig. 3, that is, perpendicular to the plane of Fig. 3. This holds for monoclinic crystals, illustrated in Fig. 4i. In general, however, a is not perpendicular to this plane; this more general relation is indicated in Fig. 4ii by the nonparallelism of a and the line normal to (100). The planes (001), (011), (021), . . . have the a axis in common, hence, are members of the same zone, and their normals lie in the same plane, which intersects the projection plane in a line somewhat

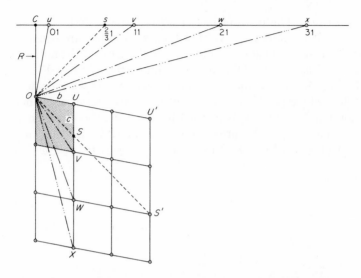

Fig. 3. The geometrical basis for the easy indexing feature of the gnomonic projection, illustrated by the indexing of rational lines in a two-dimensional lattice [from Buerger[18], p. 361 modified.]

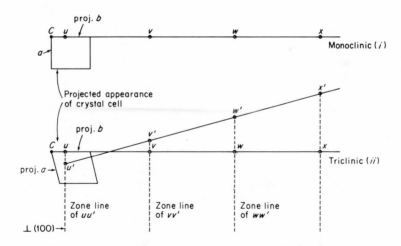

Fig. 4. The generalization of the geometry of Fig. 3 to the indexing of rational planes in a three-dimensional lattice. In each part of this illustration the view is along *OC* of Fig. 3. The polygon at the left of Fig. 4*i*, which represents the monoclinic case, is the projection along *c* of the upper left cell of Fig. 3. Figure 4*ii* represents the corresponding situation for the triclinic case.

oblique to *Cuvw*. The poles of planes (001), (011), (021), ... are u', v', w', ..., and it remains to prove that their separations are equal.

That this is so can be demonstrated by making use of the notion of zones (outlined in Chapter 2). All planes containing the axis b of Fig. 3 lie in a zone. In Fig. 4*ii* this includes a (nonrational) plane whose pole is u, the plane (001) whose pole is u', and the plane (100) whose pole is not on the projection plane because it is parallel to it, having the direction ⊥ (100). The zone line uu' on the projection plane must be parallel to the line marked ⊥ (100) in Fig. 4*ii*. In the same manner, vv' and ww' must be

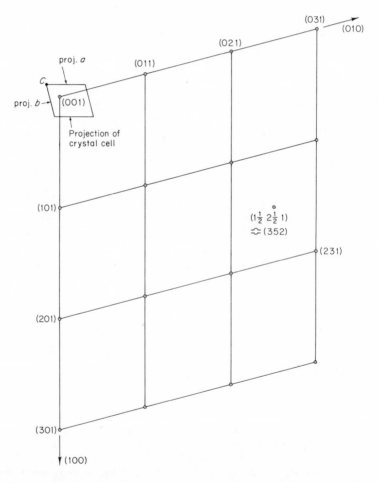

Fig. 5. View normal to the plane of the gnomonic projection, illustrating the relation between the gnomonic net and a projected cell of the crystal lattice.

parallel to this line. Thus $uu' \parallel ww' \parallel \ldots$, and since the separations in the sequence $u, v, w, \ldots$ are equal, the separations in the sequence $u', v', w', \ldots$ must also be equal (although in this general three-dimensional case the separations are larger than in the two-dimensional case, unless a is perpendicular to the plane of b and c). (The axial labels in this discussion may be interchanged in any way, provided the indices are changed in a corresponding way.)

The general consequences of this geometry with respect to the whole lattice are as follows. If a zone plane is passed normal to an axis of the crystal cell (the a axis was used in the foregoing discussion), it intersects the projection plane in a line along which the poles of the planes in that axial zone occur. The zone plane is normal to the axis, and the intersection is normal to the projection of the axis. Along the intersection the intervals between poles are equal for those planes whose third index is unity. The node where the ray from O normal to (001) reaches the projection plane may be taken as the origin node; unless a and b are perpendicular to c, this node does not coincide with the center of the projection C.

This geometry holds for any translation of the crystal lattice. In particular it holds for two axes chosen to describe the lattice, which are shown as seen looking along the third axis in Fig. 5. The normals to the faces of the crystal cell establish the directions of these line sequences. These two normals are parallel to the projection plane and so cannot intersect it to produce poles. Instead, directions parallel to the projection plane are customarily indicated by arrows, like those shown for (100) and (010) in Fig. 5.

Normal to each axis of the projected cell of the lattice there is a line of poles, as seen in Fig. 5. Along one line lie the poles of (h01), along the other the poles of (0k1). These two lines may be taken as coordinate axes of an axial system from which the rest of the network can be filled in. Each node of the network is the pole of a possible lattice plane whose third index is unity; the first two indices are readily determined from the indices of the poles along the axial lines. In Fig. 5, for example, the pole that occurs on the same net lines as (031) and (201) is (231). A pole P which occurs at the center of a cell of the network has indices ($1\frac{1}{2}$ $2\frac{1}{2}$ 1), which when multiplied through by 2 turns out to be (352).

Because only two indices are explicitly given in the gnomonic projection, these are called, generally, p and q in Victor Goldschmidt's scheme[2,11] for treating crystals. Thus the pole of the plane (352) has Goldschmidt coordinates $pq = 1\frac{1}{2}, 2\frac{1}{2}$.

Relation to the reciprocal lattice

Figure 5 resembles an oblique cell in two dimensions and its reciprocal lattice. In the illustration it is evident that the positions of the poles of the crystal planes are distributed like the points of the reciprocal lattice, that the cell of the net is geometrically the same as the reciprocal of the projection of the crystal cell, and that the axes of one cell are perpendicular to the faces of the other. That there is a relation between the gnomonic projection and the reciprocal lattice is therefore geometrically apparent.

The reason for this relation, and its limits, can be appreciated from Figs. 5 and 6. It is known from reciprocal-lattice theory that if a reciprocal lattice is constructed for a general lattice, the planes of the reciprocal lattice are perpendicular to the translations of the original lattice; in Figs. 5 and 6 the planes perpendicular to c constitute the levels of the reciprocal lattice. Each of these levels is a plane lattice. Now consider the first level (which will be seen to correspond with the gnomonic projection of the crystal): if perpendiculars are constructed to crystal planes $(hk1)$, and points placed at origin distances $1/d_{hk1}$, it is known that the resulting points fall in a plane, and the points constitute a plane lattice which is the first level of the complete reciprocal lattice. As a corollary, if perpendiculars are constructed to the crystal planes $(hk1)$, the intersection of these perpendiculars with a plane normal to c will be a plane lattice whose scale is proportional to the distance R of the plane from the origin lattice point of the crystal. The first level of the reciprocal lattice can be constructed by either method. The first method conforms to the standard way of defining the reciprocal lattice; the second conforms to the standard method of constructing the gnomonic projection of the planes $(hk1)$ of a crystal which is oriented so that its c axis is perpendicular to the projection plane. Both are the same, except for the trivial scale factor R.

The situation is different, however, for other levels. The zero level of the reciprocal lattice is parallel to the gnomonic-projection plane, but it is at a distance R below it; accordingly, no rays from O, Fig. 6, perpendicular to lattice planes of the type $(hk0)$, which are vertical, can intersect the plane of the gnomonic projection. This projection therefore contains no record of poles corresponding to points on the zero level of the reciprocal lattice.

On the other hand, points on upper levels of the reciprocal lattice do record on the gnomonic projection. But the gnomonic projection of a crystal plane is determined solely by the direction of the normal to the plane, that is, by the direction of the spacing vector **d**; wherever this vector intersects the plane of the projection, the gnomonic pole of that plane occurs. Unlike the reciprocal lattice, the reciprocal of the spacing magnitude, namely $1/d_{hk1}$, has no effect on the location of the pole. Therefore, all planes with the same spacing direction record at the same locations, and as if they were members of the first level.

This has two somewhat different results. In the first place, if one considers a reciprocal-lattice point on an upper level other than the first level, and whose three indices (or two indices, such as 23 in the two-dimensional example of Fig. 6) contain no common factor, this lattice point is the first point in the line of points radiating from the origin. Since the indices have no common factor on their own level, at least one of the first two coordinates must be fractional if the point is indexed as if it were on the first level, where the last index must be unity. The result is illustrated in Fig. 6.

On the other hand, if two of the three indices of the reciprocal-lattice point have a common factor n, then the point is the nth one along the ray from the origin; yet the poles of all planes of a set, such as 221, 442, 663, . . ., whose indices are the same except for the presence of a common

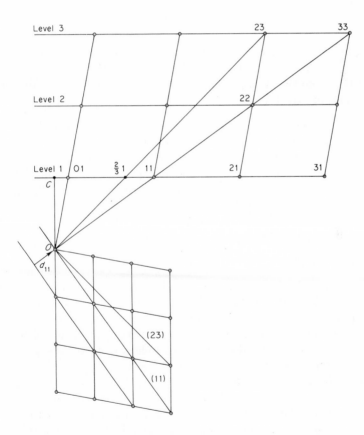

Fig. 6. Relations between the crystal lattice (below), the reciprocal lattice (above), and the gnomonic projection.

factor, record as if they were on the first level. All these reciprocal-lattice points therefore correspond to the one pole on the gnomonic projection; in the example, all the reciprocal-lattice points are represented by the pole of (221), that is, at $pq = 11$. For the study of surface planes, or of stacks of rational planes, planes whose indices have a common factor are meaningless, hence, that feature of the gnomonic projection brings no disadvantage. But when two planes with indices differing by a common factor have different meanings, as they do in the reflection notation of x-ray diffraction, this feature is a serious disadvantage. This matter is developed further in connection with the indexing of Laue photographs in Chapter 5.

Basic constructions

As in the case of stereographic projections, the use of the gnomonic projection can proceed by using only the basic tools of geometry: a pencil, straightedge, compass, and protractor. With these tools, the fundamental constructions are the following[2,11].

a. To plot a pole, given its polar angle. This is the fundamental construction in the gnomonic projection which was illustrated in Fig. 1. The meridian plane in which the pole lies intersects the gnomonic-projection plane in a line Cp. If the reference sphere is imagined rotated 90° about this line, the geometry would appear as in Fig. 1. Line OC is then the direction of the north pole, so that the polar angle ρ is measured off from this line, thus determining line OPp, intersecting the profile of the projection plane at p. When the sphere is rotated back to its initial position, p is the required gnomonic pole of P.

The angular coordinates ρ and ϕ of the pole of a crystal face are most conveniently established with the aid of a two-circle goniometer. The general principles are suggested in Figs. 7 and 8. Figure 7 shows a meridian plane making an angle ϕ with a prime meridian; in this plane lies a pole P which makes an angle ρ with the radius to the north pole of the sphere. The usual orientation of the Goldschmidt two-circle goniometer is suggested in Fig. 8i. If the measuring circles of the goniometer are to have the same orientation as the arcs of ρ and ϕ in Fig. 7, the goniometer must be reoriented as in Fig. 8ii, so that its normally vertical (V) circle becomes fixed in a horizontal plane, while the rest of the goniometer swings about the axis of this circle, thus making the normally horizontal (H) circle vertical. To determine the values of ρ and ϕ, an arrangement is used which makes use of an optical reflection in order to locate the normal to a crystal plane, such as plane F in Fig. 8ii. The simplest optical arrangement is that of an auto-

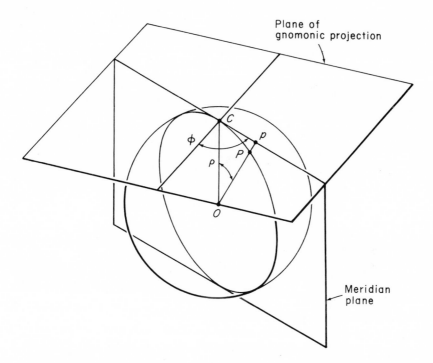

Fig. 7. The spherical coordinates ρ and ϕ of a crystal face, the pole P of the face on the reference sphere, and its gnomonic projection p.

collimator. This is a device which sends a light beam along its axis toward the crystal; if the normal to the crystal face is made to coincide with the axis of this light beam, the beam is reflected back into the autocollimator, where it can be seen by the eye. The arc adjustments that must be made to cause this condition to occur are the required values of ϕ and ρ for the crystal face.

b. To plot the zone line of any two poles. This construction is trivial in the gnomonic projection because every zone projects as a straight line. Given two poles, therefore, their common zone is the straight line drawn through them. (The zone line was a great circle in the stereographic projection.)

c. To find the pole ("*edge pole*") *of a zone.* In Fig. 9, the zone line gg' on the gnomonic-projection plane is given, and it is required to find the pole of the zone in the projection. It should first be noted that the line through the midpoint C of the projection, and normal to the zone line, is called the *central* of the zone. The intersection n of the central and its zone is called the *zone center*.

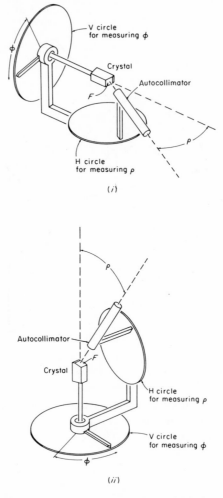

Fig. 8. A two-circle goniometer for measuring spherical coordinates ρ and ϕ. (*i*) The instrument is seen in its true position. (*ii*) The instrument reoriented to correspond with Fig. 7.

The zone line gg' is the gnomonic projection of the great circle GG' on the sphere. If the sphere is imagined rotated 90° about the central, so that the radius OC comes to lie in the plane of the paper, as in Fig. 10, this plane becomes the meridian plane containing the central. The central is not affected by the rotation. The line normal to nO at O, namely OZ, locates the pole Z on the sphere, and if extended till it intersects the central, determines the required gnomonic pole z. The actual construction is shown in Fig. 11. Given the zone line gg', its central Cm is first drawn, intersecting the zone line at n. At C, a circle of radius R (the radius of the reference sphere) is drawn, and a line fCf, normal to the central, is also drawn. At the intersection o of fCf and the circle, a line is drawn perpendicular to no. The intersection of this line with the central determines the desired gnomonic pole z.

d. To find the zone line of an edge pole. This construction is the reverse of the preceding one. In Fig. 9, edge pole z is given, and the zone line gg' corresponding to it is required. The construction is that of Fig. 11. A line zC is first drawn. This line, extended, is the central zCm of the required zone. At C a line fCf is drawn perpendicular to the central, and at C a circle of radius R (the radius of the reference sphere) is also drawn. The intersection of fCf and the circle locates o. Line zo is drawn, then line on at right angles to it, thus locating n. The line through n perpendicular to the central is gng', the required zone line.

e. To find the angle between two poles. The angle between the two poles P and Q on the reference sphere, Fig. 12, is required. Their gnomonic projections are p and q. These determine a zone line $gpqg'$ whose central is Cnm. The angle between P and Q is ψ, the angle they subtend at the sphere's

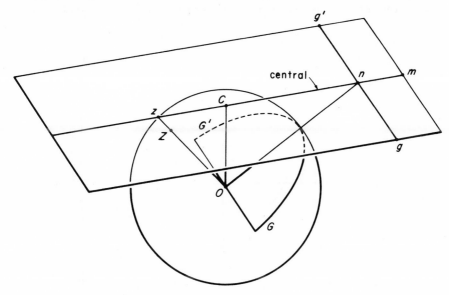

Fig. 9

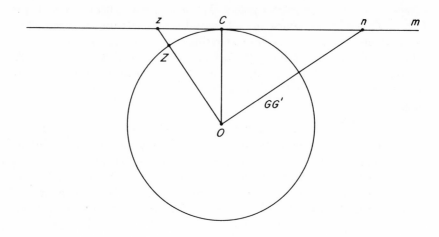

Fig. 10

center O. Suppose the triangle nOp is rotated about the zone line until O migrates to w in the plane of projection. Then the $\angle pwq = \angle pOq = \psi$, the required angle. It remains to locate w on the central Cnm. If the meridian plane containing the central is rotated 90° about the central, so that it lies in the plane of the paper, the view is that of Fig. 13. There it is seen

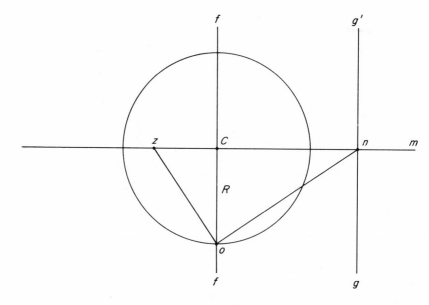

Fig. 11

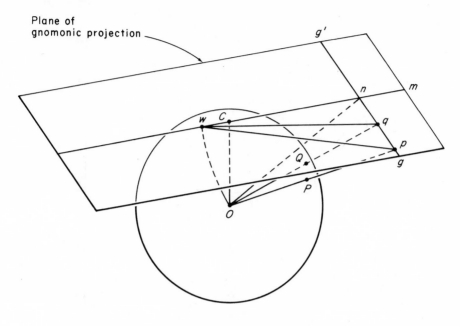

Fig. 12

that if an arc of radius nO is swung about n, it intersects the central Cn at w. The actual construction is carried out as outlined in Fig. 14. The required angle between poles P and Q is $\psi = \angle pwq$. The point w is called the *angle point* of the zone.

f. To locate a pole in a zone line, given its angle to another pole in the zone.

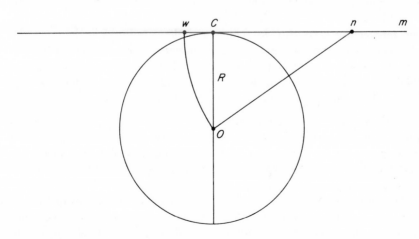

Fig. 13

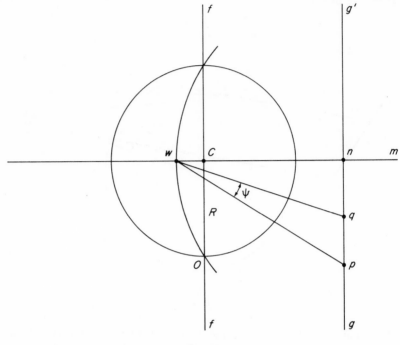

Fig. 14

This is a corollary to the last construction. In Fig. 12, the zone line gg' is given, and in it a pole p. It is required to find a pole q in the same zone and making an angle ψ with p. The construction is carried out as indicated in Fig. 14. The central Cnm is first constructed to the given zone, thus locating the zone center n. With C as center, a circle of radius R is drawn, and a perpendicular to the central at C is also drawn. The intersection of this line and the circle locates O. With n as center and nO as radius, an arc is drawn; its intersection with the central locates the angle point w. Angle pwq equal to ψ is then laid off at w, thus locating the desired pole q.

g. To find the angle between two zones. The geometry involved in this problem is illustrated in Fig. 15, which is a view of the reference sphere looking toward its north pole, that is, looking normal to the projection plane. In Fig. 16 the corresponding gnomonic projection, in the same orientation, is shown.

The angle required is the dihedral angle between the planes of the two great circles G_1G_1' and G_2G_2' in Fig. 15. This is the same angle (or its supplement) as the angle between the normals Z_1 and Z_2 to these planes. The problem of finding the angles between these two (edge) poles in the gnomonic projection was solved in construction e. A shortcut in the construction, however, is available from the following considerations.

The pole Z_1 is perpendicular to plane G_1G_1';
the pole Z_2 is perpendicular to plane G_2G_2';
$\therefore$ the plane of Z_1Z_2 is perpendicular to the intersection of G_1G_1' and G_2G_2', namely, the line OP;
$\therefore$ OP is the pole of zone Z_1Z_2.

The meridian section through P is the plane POM. It intersects zone Z_1Z_2 at N_3, whose gnomonic projection in Fig. 16 is n_3.

The straightforward (but longer) way of performing this construction is as follows. In Fig. 16 the zone lines g_1g_1' and g_2g_2' are given. Using construction c, the pole z_1 of zone g_1g_1' and the pole z_2 of zone g_2g_2' are first found. It is the angle between these poles (or their supplement) which is required. To carry this out, z_1 and z_2 are treated like poles in a zone, so that the angle point of this zone is sought. To find this, the line z_1z_2, and then the central n_3C, are drawn. Along this central the angle point w_3 is located by construction e.

The shortcut is that z_1 and z_2 need not be located separately by finding their positions along the centrals of their individual zones. Rather, the

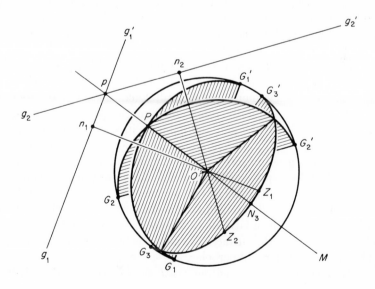

Fig. 15

original zones g_1g_1' and g_2g_2' intersect at p, which determines the meridian line pCm. On this line, at an angular distance of 90° from p, occurs n_3, which is the center of the required zone g_3g_3'. Thus, z_1 and z_2 may be located at the same time on the centrals of their individual zones by drawing zone g_3g_3' at right angles to pCm at n_3.

Aids in using the gnomonic projection

Scales and "protractors". The simplest aid for plotting the gnomonic projection of a pole is a scale[8] which provides the central distance r in (1) as a function of the coordinate ρ. The preparation of such a scale requires a decision concerning the magnitude of the radius R of the sphere. This is commonly chosen as 5 cm. Such a scale is ordinarily a part of a stereographic protractor mentioned in Chapter 2.

The gnomonic nets. Just as latitude and longitude lines can be mapped from the globe onto the stereographic projection, so they can be mapped onto the gnomonic projection. When the gnomonic projection is made on a plane tangent to the globe at the north pole, the resulting net, called the *polar gnomonic net*, is as shown in Fig. 17. The meridians appear as they do on any perspective projection, namely, as radiating straight lines with true angular separation. The parallels, however, are small circles coaxial with the south–north axis. They project as concentric circles with radii given by (1). This net provides coordinates ρ and ϕ, which may be used to plot the poles of faces measured with a two-circle goniometer. When ρ exceeds about 75°, however, it becomes impracticable to plot a pole; this is one of the disadvantages of the gnomonic projection[†]. Faces in the prism zone, for which $\rho = 90°$, are ordinarily represented in the projection by arrows drawn around the rim of the plot and pointing in the direction of the face normal.

Figure 18 shows a net of latitude and longitude lines as projected on a plane tangent at the equator. The meridian lines are projections of great circles on the sphere, having the common diameter of the north–south axis. They are straight lines, all parallel to the north–south axis, in the gnomonic projection. The parallels of latitude on the sphere project to the gnomonic-projection plane along the surfaces of cones coaxial with the north–south axis. They are accordingly hyperbolas.

This net, called the *equatorial gnomonic net*, was first presented by Hilton[7];

[†] A way of avoiding this disadvantage is treated in Chapter 4.

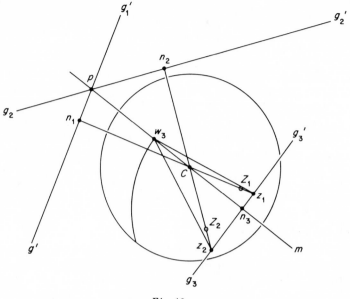

Fig. 16

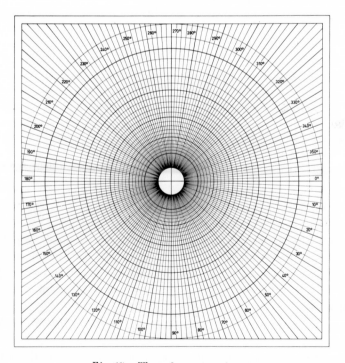

Fig. 17. The polar gnomonic net.

73

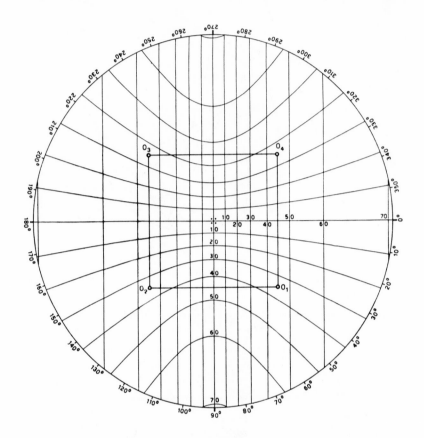

Fig. 18. The equatorial (meridional or Hilton) gnomonic net. [From Terpstra and Codd[20], p. 46.]

it serves the same purpose in the gnomonic projection as the Wulff net does in the stereographic projection: it permits determining what happens to points in the gnomonic projection when (i) the gnomonic projection is rotated about its center, or (ii) the reference sphere is imagined rotated about its north–south axis. For these operations the Hilton net is manipulated exactly like the Wulff net of the stereographic projection, as discussed in Chapter 2.

Crystal drawing

As was pointed out in the discussion of crystal drawing with the aid of the stereographic projection, a crystal drawing consists basically of straight lines, each having the direction of an edge between two intersecting crystal faces. These edge directions can also be determined from the gnomonic projection.[2,6,12-15,20] As a simple example of the basic principle, consider the relation of crystal edges as seen when looking at the crystal in a direction normal to its gnomonic projection. This would give a top view of the crystal. As can be seen in Fig. 19, for any zone GG', there is a normal OZ which is the actual direction of the edge between any two planes in the zone. As seen looking at the crystal in a direction normal to the plane of the gnomonic projection, this edge appears projected as the line Cz. This direction is that of the central of the zone line gg'. In a view looking normal to the gnomonic projection, therefore, the direction of an edge between two faces is normal to the zone line in which the two faces lie.

This simple relation holds when the viewing direction is represented by the center of the gnomonic projection. If another viewing direction is required, this relation can also be used if the reference sphere, together with its face poles and desired viewing direction, is rotated till the viewing direction occupies the center of the gnomonic projection. This is ordinarily an undesirable procedure.

The crystal can also be drawn as seen from any viewing direction without rotating the reference sphere, by using a more complicated procedure based

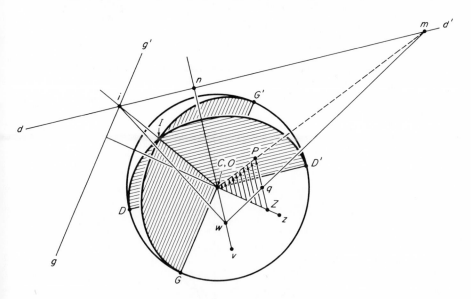

Fig. 19. Geometrical relations leading to determining the direction of an edge between two faces with the aid of the gnomonic projection when making a drawing of a crystal.

on the following geometry. In Fig. 19, let the zone plane whose edge is
required be GG' (projecting as gg') and the plane on which the drawing is
to be made be DD' (projecting as dd'). These two planes intersect along
OI (projecting as i). The actual edge direction, that is, the normal to plane
GG', is OZ. This line is to be viewed normal to plane DD', however, so that
it appears to the eye as its projection on DD', namely OP. Then,

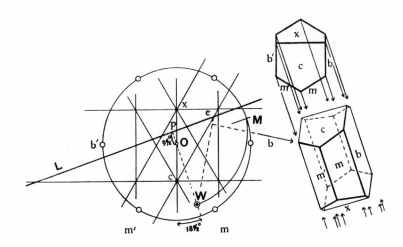

Fig. 20. Drawings of a crystal of orthoclase made with the aid of the gnomonic
projection. [From Barker[14], p. 56.]

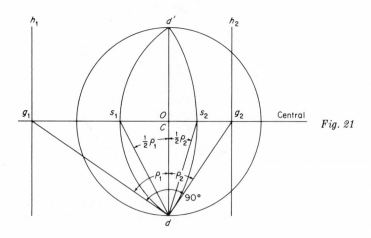

Fig. 21

Plane $DD' \perp$ line PZ,

Plane $GG' \perp$ line OZ,

∴ Intersection line $OIi \perp$ plane POZ,

∴ Intersection line $OIi \perp$ line OP.

The direction OP of the edge is now known in the plane normal to the viewing direction. But the actual drawing will be made on the paper, which is the plane of the gnomonic projection, so the direction of OP on the gnomonic projection is required. This can be found by rotating the drawing plane $DD'dd'$ about dd' into the gnomonic-projection plane. In this process, points on the line dd' such as intersection i, m, and n, the central of dd, remain invariant, while $O \rightarrow w$, $P \rightarrow q$, and $\angle iOP \rightarrow \angle iwq$, which remains $90°$, where w is the angle point of line dd'. The line wq, which is perpendicular to iw, is the direction of the required edge.

The line dd', which represents the intersection of the drawing plane with the gnomonic-projection plane, is called the *guide line*. To draw the edge between two faces in a zone, first find the intersection i of the zone with the guide line. Then connect i with the angle point w of the guide line, and lay off angle $iwq = 90°$. This is the required edge direction.

A simple example of a crystal drawn by this method is shown in Fig. 20. The construction used in finding the direction of one edge is also shown in the illustration.

Some relations between the gnomonic and stereographic projections

Some simple relations exist between the gnomonic and stereographic projections.[2,11] These are basically dependent on the relation between the aperture angles subtended by a point and the north pole, as seen from the projection points, in the two projections. In both projections, the plane of the projection is at a distance equal to the radius R of the sphere from the projection center. If the polar distance of a point on the sphere is ρ, its aperture angle from the north pole is ρ in the gnomonic projection, but $\frac{1}{2}\rho$ in the stereographic projection.

In Fig. 21, the circle represents the profile of the reference sphere as seen looking toward the north pole. The horizontal line is a central of some zone in both stereographic and gnomonic projections. Let g_1 and g_2 be the gnomonic projections of two points $90°$ apart on the reference sphere. Victor Goldschmidt called such a pair of points *conjugate points* because if one is an edge pole, the other is the center of the zone corresponding to it. Thus the zone line corresponding to edge pole g_1 is g_2h_2, and the zone line corresponding to edge pole g_2 is g_1h_1.

Suppose that g_1 subtends aperture angle ρ_1 at the gnomonic-projection center. The corresponding aperture at the stereographic-projection center is $\frac{1}{2}\rho_1$. Thus, s_1 and s_2 are conjugate points in the stereographic projection corresponding to g_1 and g_2 in the gnomonic projection.

Now, if g_1 is the center of a zone, the zone line is g_1h_1; the corresponding stereographic zone line is a great circle through s_1 which intersects the fundamental circle at the ends of diameter dd'. That is, the stereographic zone circle ds_1d' corresponds to the gnomonic zone line g_1h_1. (Similarly, stereographic zone circle ds_2d' corresponds to gnomonic zone line g_2h_2.)

Some of the angular relations in Fig. 21 are shown separately in Fig. 22. There it is seen that because g_1 and g_2 are conjugate,

$$\angle g_1dg_2 = 90°$$

and

$$\angle g_1dO = \rho_1.$$

Since stereographic s_1 corresponds to gnomonic g_1,

$$\angle s_1dO = \tfrac{1}{2}\rho_1.$$

Further,

$$\angle g_1ds_1 = \angle g_1dO - \angle s_1dO = \rho_1 - \tfrac{1}{2}\rho_1 = \tfrac{1}{2}\rho_1$$

$$\angle s_1dg_2 = \angle g_1dg_2 - \angle g_1ds_1 = 90° - \tfrac{1}{2}\rho_1.$$

In right triangle s_1Od

$$\angle ds_1O = 90° - \tfrac{1}{2}\rho_1.$$

$\therefore$ s_1dg_2 is isosceles,

$\therefore$ $s_1g_2 = dg_2$,

$\therefore$ in Fig. 21, the center of great circle ds_1d' is g_2.

In the same manner, it can be shown that the center of great circle ds_2d' is g_1.

Theorem. The gnomonic edge pole of a zone (for example, g_1) is the center of the arc of the stereographic zone circle (in the example, ds_2d').

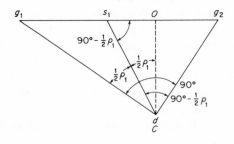

Fig. 22

But g_1 is separated from g_2 by $90°$, so that g_1 is the angle point of zone line g_2h_2. This provides the further relation

Theorem. The gnomonic angle point of a zone (for example, zone g_2h_2) is the stereographic edge pole (in this example, s_1).

These theorems make it possible to transform from one projection to another, zone by zone.

Summary

The gnomonic projection is one of the perspective projections. The points on the surface of the reference sphere are projected from the sphere's center to a tangent plane; the point of tangency is ordinarily the north pole.

The great advantage of this projection in crystallography is that if an axis of the crystal is oriented normal to the projection plane, the poles of the lattice planes fall on the nodes of a main net in the projection, or on the nodes of subsidiary nets bearing subcell relations to that of the main net. Thus the locations of the poles on the gnomonic projection bear a geometrical relation to locations of some of the points on the reciprocal lattice. If the c axis is oriented normal to the projection plane, poles corresponding to $hk0$ points do not record on the projection, while poles of the $hk1$ planes fall on points of a main net which has the geometry of the first level of the reciprocal lattice. All planes whose indices are the same except for different common factors have the same gnomonic-projection poles. Planes whose indices do not have a common factor, and for which $l > 1$, have poles which fall at points that have one or two fractional coordinates with respect to the main net.

In using the gnomonic projection, the constructions and manipulations may be carried out with the simple tools ordinarily used in geometry. Alternatively, a coordinate system of latitude and longitude lines provided by a network known as the *Hilton net* may be used.

As in the case of the stereographic projection, the gnomonic projection can also be used in the drawing of crystals. The trace of the plane normal to the direction chosen for viewing the crystal is called the *guide line*. With its aid, the direction of the various edges of the crystal can be determined for the viewing direction selected.

The gnomonic and stereographic projections have interesting interrelations which are useful in transforming from one to another: the gnomonic angle point of a zone is the stereographic pole of that zone, and a gnomonic edge pole is the center of the stereographic zone circle of that edge pole.

History

Although the gnomonic projection may have been known to Ptolemy, the first written record of it was published in Rome in 1612 by Christopher Grienberger in his *Prospectiva nova coelestis*. Accordingly, the first use of this projection was for mapping star locations in the sky. It was later used in navigation, and maps using gnomonic projections are still in use.

The application of the projection in crystallography was apparently first made by Mallard[1], who not only gave illustrations of the properties of planes and zones for various crystal systems, but also recognized the relation of the gnomonic projection to the polar lattice, now known as the reciprocal lattice. The gnomonic projection was also adopted about the same time by Victor Goldschmidt[2] and his students. In papers published in the United States up to about the second world war, it became a standard scheme for representing observed crystal forms. At about that time the study of crystal faces declined as the investigation of crystals by x-ray diffraction methods swept the country. The gnomonic projection has maintained its popularity because it proves useful for indexing the reflections on Laue photographs. Most other methods of x-ray diffraction make use of the reciprocal lattice, to which the gnomonic projection is closely related.

Significant literature

1. Ernest Mallard. Traité de cristallographie géométrique et physique. (Dunod, Paris) Vol. 1 (1879) 372 pages, especially 63–66. Atlas (1879) 9 plates, of which the first 6 are gnomonic projections.
2. Victor Goldschmidt. Ueber Projection und graphische Krystallberechnung. (Springer, Berlin, 1887) 97 pages.
3. H. A. Miers. On the use of the gnomonic projection; with a projection of the forms of red silver ore. *Mineralog. Mag.* **7** (1887) 145–149.
4. V. Goldschmidt. Graphische Bestimmung des Winkels zweier Zonenebenen in gnomonischer Projection. *Z. Kristallogr.* **17** (1890) 97.
5. V. Goldschmidt. Zur graphischen Kristallberechnung. *Z. Kristallogr.* **20** (1892) 143–145.
6. G. F. Herbert Smith. On the advantages of the gnomonic projection and its use in the drawing of crystals. *Mineralog. Mag.* **13** (1903) 309–321.
7. Harold Hilton. The gnomonic net. *Mineralog. Mag.* **14** (1904) 18–20.
8. A. Hutchinson. On a protractor for use in constructing stereographic and gnomonic projections of the sphere. *Mineralog. Mag.* **15** (1908) 93–112, especially *Historical appendix*: 105–112.
9. H. E. Boeke. Die gnomonische Projektion in ihrer Anwendung auf kristallographische Aufgaben. (Borntraeger, Berlin, 1913) 54 pages.
10. Charles Palache. The Goldschmidt two-circle goniometer. *Amer. Mineralogist* **5** (1920) 23–33.

11. Charles Palache. The gnomonic projection. *Amer. Mineralogist* **5** (1920) 67–80.
12. Mary W. Porter. Practical crystal drawing. *Amer. Mineralogist* **5** (1920) 89–95.
13. Charles Palache. Further notes on crystal drawing. *Amer. Mineralogist* **5** (1920) 96–99.
14. T. V. Barker. Graphical and tabular methods in crystallography. (Thomas Murby & Co., London, 1922) 152 pages.
15. Robert L. Parker. Kristallzeichnen. (Gebrüder Borntraeger, Berlin, 1929) 112 pages [with 15 literature citations on crystal drawing].
16. F. E. Wright. Shift of the plane of projection in the gnomonic projection. *Amer. Mineralogist* **17** (1932) 423–428.
17. Victor Goldschmidt. Kursus der Kristallometrie. [edited by Hans Himmel and Karl Müller]. (Gebrüder Borntraeger, Berlin, 1934) 167 pages.
18. M. J. Buerger. Lattice indices and transformations in the gnomonic projection. *Amer. Mineralogist* **19** (1934) 360–369.
19. D. W. Dijkstra. Transformation of gnomograms and its application to the microchemical identification of crystals. *Proc. Kon. Ned. Akad. Wet.* **52** (1949) 3–18.
20. P. Terpstra and L. W. Codd. Crystallometry (Longmans Green, London, 1961) 420 pages, especially 3–11.
21. R. L. Cunningham and Joyce Ng-Yelim. An optical device for the direct production and viewing of stereographic and gnomonic projections. *J. Appl. Cryst.* **1** (1968) 320–321.

Chapter 4

The stereognomonic projection

Background

Since the stereographic and gnomonic projections are both perspective projections, it is natural that there should exist relations between them. Some of these were mentioned at the end of Chapter 3. Unfortunately, those relations, though interesting, are not very useful. It remained for Robert L. Parker[1–3] to formulate another relation between these two projections; this formulation led him to a new transformation from one projection to the other and, by making use of it, to a most useful combination of the two projections which preserves the best features of both.

As applied in crystallography, the gnomonic projection has the advantage of placing the poles of rational crystallographic planes on the nodes of a network that permits indexing of the planes by inspection. It has the serious disadvantage of being unable to display within the area of a reasonably large sheet of paper any poles whose ρ coordinate is larger than about 70°: this practical disadvantage turns into the theoretical impossibility of locating poles whose ρ coordinates are 90°, which applies to all planes in the prism zone. The stereographic projection, on the other hand, neatly represents such planes on the fundamental circle, but it suffers from the disadvantage that planes cannot be indexed by inspection of the locations of their poles. Parker's combination projection retains the advantages of both projections and avoids their disadvantages.

Transformation between stereographic and gnomonic projections

Traditional transformation. A specific pole of a crystal can be located on the spherical projection by spherical coordinates ρ (the angular distance from the north pole of the spherical projection) and ϕ, the meridian coordinate. The meridian coordinate of the pole is the same when projected on the stereographic or the gnomonic projection, so the projections differ only in the treatment of the angle ρ.

Figure 1 illustrates the obvious and usual way of transforming a point from the stereographic to the gnomonic projection, or the reverse. A view of the sphere normal to the meridian circle containing the pole P is shown in Fig. 1i. The planes of both the stereographic and gnomonic projections are at the same distance R from their projection points. It is convenient to use subscripts s and g to represent the stereographic and gnomonic projections, respectively. The central distance of the stereographic projection of pole P is

$$Op_s = R \tan \tfrac{1}{2}\rho, \tag{1}$$

while the central distance of the gnomonic projection of pole P is

$$Op_g = R \tan \rho. \tag{2}$$

Thus the ratio of the central distances is

$$\frac{Op_s}{Op_g} = \frac{\tan \tfrac{1}{2}\rho}{\tan \rho}. \tag{3}$$

The construction corresponding to this simple theory is shown in Fig. 1ii, which can be regarded as a view of the stereographic projection. The meridian plane containing the stereographic pole p_s and the gnomonic pole p_g is seen on edge as line CM. Diameter DD' is drawn through C at right angles to the line of the meridian plane. Angles $\tfrac{1}{2}\rho$ and ρ are now laid out from line DC at D to locate p_s and p_g. In transforming from stereographic to gnomonic projections, the angle CDp_s is doubled to locate p_g; in transforming from gnomonic to stereographic projections, the angle CDp_g is halved to locate p_s.

The circle of concordance. Parker's approach[1,3] was characterized by permitting the radius R of the sphere to be a variable. Then the central distances of a pole on the stereographic and gnomonic projections are

$$\text{stereographic:} \qquad d_s = R_s \tan \tfrac{1}{2}\rho, \tag{4}$$

$$\text{gnomonic:} \qquad d_g = R_g \tan \rho. \tag{5}$$

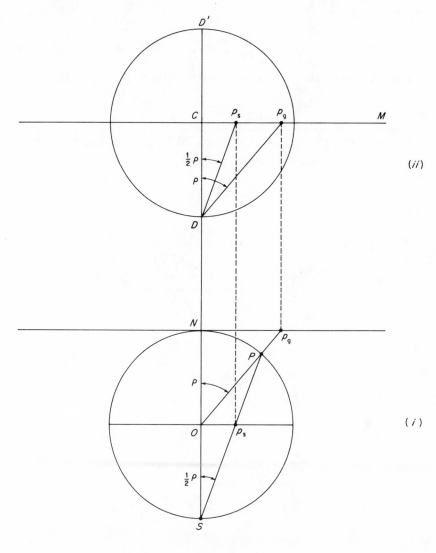

Fig. 1

If d_s is set equal to d_g, there results

$$R_s \tan \tfrac{1}{2}\rho = R_g \tan \rho. \qquad (6)$$

This is an equation in the three variables R_s, R_g, and ρ. If any two are fixed, the third is determined. In particular, it is of interest to rearrange

(6) in the form

$$\frac{\tan \rho}{\tan \frac{1}{2}\rho} = \frac{R_\text{s}}{R_\text{g}} \tag{7}$$

and explore some values of ρ in terms of the ratio R_s/R_g. A few such values are

R_s/R_g	3	10/3	4	5	∞
ρ	60.00°	64.42°	70.53° ($70°31\frac{1}{2}'$)	75.52°	90°

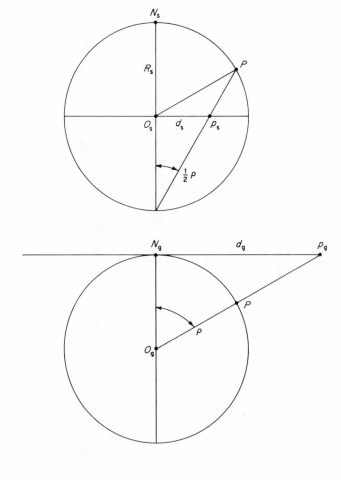

Fig. 2

The value of $R_s/R_g = 4$ corresponds to a value of ρ in the neighborhood of the region where the gnomonic projection became unwieldy.

This discussion is illustrated in Figs. 2 and 3. Figure 2 shows the relation between stereographic and gnomonic distances d as a function of angle ρ. Figure 3 shows how a particular point P can be given the same location p on both stereographic and gnomonic projections by adjusting the radius ratio R_s/R_g. It must be emphasized that when it is decided to make the stereographic point p_s and the gnomonic point p_g coincide, this coincidence occurs only for the particular value of ρ determined by the ratio chosen in (7). Only such points p as have this particular value of d in Fig. 3 coincide. Such points lie on a small circle of radius d in both stereographic and gnomonic projections. This is called the *circle of concordance* of the two projections.

Relations within the circle of concordance. If the circle of concordance has been chosen so that it represents the outer limit of convenient use of the gnomonic projection, then a relation between stereographic and gnomonic projections beyond this circle has little meaning except for points on the primitive circle of the stereographic projection. These points are the transforms of points represented by horizontal arrows in the gnomonic projection.

Within the circle of concordance there is an interesting and simple relation between the projection of a point P on the sphere when projected stereographically as p_s and as projected gnomonically as p_g, illustrated in Fig. 4. Both projected points lie on the same meridian with the same coordinate ϕ. The stereographic point p_s is on a great circle ap_sb which is

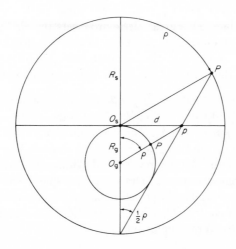

Fig. 3

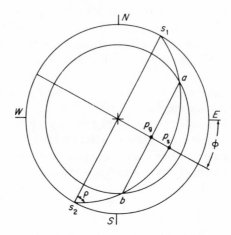

Fig. 4. Relations between stereographic pole p_s and gnomonic pole p_g within the circle of concordance. [From Parker[1], p. 498, with additional symbols.]

inclined at an angle ρ to a vertical great circle perpendicular to the meridian, as seen in Fig. 4. This great circle ap_sb cuts the circle of concordance at a and b, so that the stereographic circular line ap_sb must correspond with gnomonic straight line ab. Since p_g is on the same meridian as p_s, the intersection of ab locates the gnomonic point p_g.

Construction of a gnomonic point from a stereographic point. With the radius ratio R_s/R_g adjusted to a circle of concordance, draw a meridian line through stereographic point p_s and draw a great circle through p_s normal to this meridian line. This great circle intersects the circle of concordance at two points (a and b, Fig. 4); connect these two points by a straight line (which should be normal to the meridian line). The intersection of this straight line and the meridian line is the corresponding gnomonic point p_g.

Construction of a stereographic point from a gnomonic point. With the radius ratio R_s/R_g adjusted to a circle of concordance, draw a meridian line through the gnomonic point p_g. The perpendicular to this line at p_g intersects the circle of concordance at two points (a and b, Fig. 4). Construct a great circle containing the two points, and symmetrical with the meridian. (This can be done as follows. Draw straight line s_1a and construct its perpendicular bisector. The intersection of the bisector and the meridian is the center of great circle s_1abs_2.) The intersection of this great circle and the meridian line is the stereographic point p_s.

The combination of stereographic and gnomonic projections

Advantages of the stereognomonic projection. Although it is philosophically satisfying to be able to transform at will from stereographic to gnomonic projection, or the reverse, the great practical advantage of Parker's theory of the relation between these two projections is that it may be utilized to construct a combination of the two projections, called the *stereognomonic[†] projection*[1–4]. This projection is pure gnomonic within the circle of concordance, and pure stereographic outside the circle of concordance. Examples of this type of projection are given in Figs. 5 and 6. Inspection of these illustrations makes the advantages of the stereognomonic projection obvious.

In the first place, the central (gnomonic) part of the projection permits establishing the indices of the poles of the crystal by the way they fall on the nodes of the gnomonic net. Each straight line on this net is a zone line. Each straight line extends from one part of the circumference of the circle of concordance to another part. Beyond the circle of concordance, the straight line immediately transforms into the corresponding great-circle zone line of the stereographic projection. Thus, a zone line is a straight line within the circle of concordance and becomes continuous with two parts

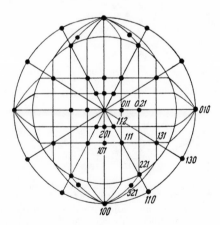

Fig. 5. Stereognomonic projection of the form development of a topaz crystal. [From Parker[1], p. 498.]

[†] The combination was first called a "gnomono-stereogram" by Parker[1,2]; later[3] he called the net a "stereo-gnomonic" net. Terpstra and Codd[4] call the projection a "gnomo stereographic" projection. We prefer "stereognomonic" as the adjective.

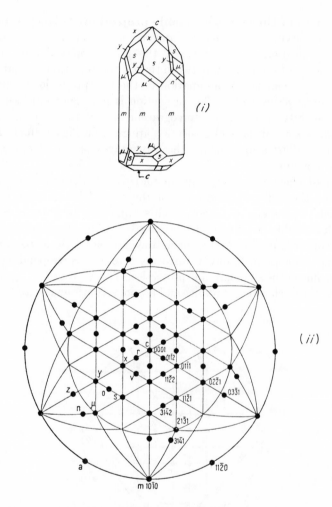

Fig. 6. (*i*). Form development of a crystal of apatite from Val Casatscha. [From Parker[2], p. 500.] (*ii*). Stereognomonic projection of form development of crystal shown in Fig. 6*i*. Letter near a pole projection corresponds with a face with the same letter in Fig. 6*i*. [From Parker[2], p. 499.]

of a great circle (each part symmetrical with the normal to the straight line) outside the circle of concordance. In this way, the terminal faces of a crystal ordinarily fall within the circle of concordance and are identified in the gnomonic part, while the prism faces, and some few others, fall outside the circle of concordance.

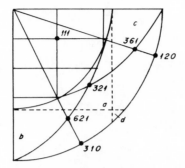

Fig. 7. Fields of the stereognomonic projection. [From Parker[1], p. 499.]

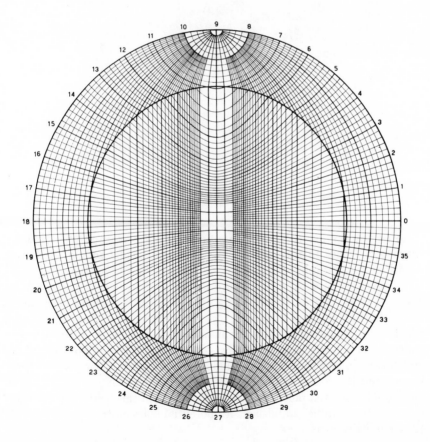

Fig. 8. Parker's meridional stereognomonic net. [From Parker[3], p. 594.]

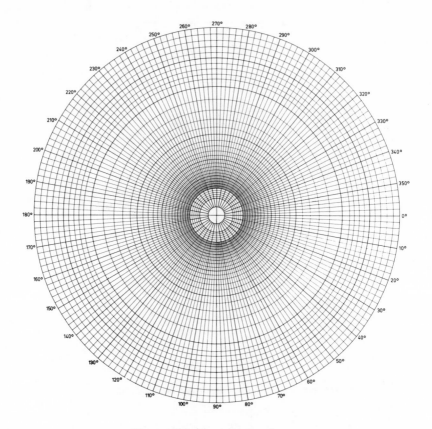

Fig. 9.   Polar stereognomonic net.

Some properties of the various fields. Parker pointed out[1] that by drawing tangents to the circle of concordance parallel to the axial zone lines, the stereographic part of the stereognomonic projection is divided into four fields, labeled *a*, *b*, *c*, and *d* in Fig. 7. Provided that a crystallographic axis is vertical, the following discussion holds for indexing poles in the various fields. In field *a*, every potential pole is on two zone circles which are continuations of two gnomonic zone lines, and so can be indexed. In fields *b* and *c* every potential pole is on one zone circle and one meridian, and can be indexed, for the meridian gives the ratio of two of the three indices and the zone circle establishes the absolute value of one of them. In field *d* a potential pole is on a meridian only; in this field, only the ratio of two indices can be established. All potential poles on the primitive circle have one index zero.

Some practical considerations. Parker supplied a net[3] for plotting stereognomonic projections. This net is reproduced on half scale in Fig. 8. It is based upon a radius ratio $R_s/R_g = 4$ for which $\rho = 70°31\frac{1}{2}'$. The original net is for $R_s = 10$ cm, which is a standard for which stereographic nets are available. Outside the circle of concordance the net is Wulff's meridional stereographic net. Within the circle of concordance it is Hilton's meridional gnomonic net. A polar stereognomonic net that matches Parker's meridional net of Fig. 8 is provided in Fig. 9.

Significant literature

1. Robert L. Parker. Eine praktische Methode zur Transformation von Kristallstereogrammen in gnomonische Projektion. *Zeit. für angew. Math. u. Physik* **5** (1953) 497–499.
2. Robert L. Parker. Über Apatit von Val Casatscha. *Schweiz. Min. Petr. Mitt.* **34** (1954) 498–501.
3. Robert L. Parker. Ein neuartiges stereo-gnomonisches Netz. *Schweiz. Min. Petr. Mitt.* **36** (1956) 593–594.
4. P. Terpstra and L. W. Codd. Crystallometry. (Longmans, Green, London, 1961) 420 pages, especially 47–50.

Chapter 5

The conventional Laue method

Experimental conditions

Radiation. The Laue method was the vehicle of discovery of the diffraction of x rays by crystals; hence, it is the oldest x-ray diffraction method. The characteristic feature of the Laue method is that it makes use of a crystal which remains fixed with respect to the x-ray beam. In the original and most common present-day applications of the method, it is important to have x radiation in which the wavelength is a continuous variable over a reasonably large range. Ideally, this requirement calls for taking the x radiation from a tube having a heavy-metal target, commonly tungsten, because such a tube can be operated at a fairly high potential without producing characteristic radiation. For many purposes, however, the presence of characteristic radiation along with the general radiation is not objectionable, so that unfiltered radiation from an x-ray tube having a molybdenum target (available in most laboratories for providing characteristic $MoK\alpha$ radiation of wavelength $\lambda = 0.7101$ Å) is often used.

X radiation having a continuously variable wavelength is not always a vital feature of the Laue method. Indeed, in the study of diffuse reflections, such radiation plays no substantial role; instead, the characteristic component is the required radiation. This is because a fixed crystal requires a continuous variation of either λ or σ ($\eqsim 1/d$) for diffraction. The latter variable is a characteristic of diffuse reflections, the investigation of which is discussed in Chapter 12.

In certain applications of the Laue method the range of wavelengths used must be known. The minimum wavelength depends on the voltage V across the x-ray tube according to the relation

$$\lambda_{\min}V = \frac{hc}{e},$$ (1)

$$= 12.398 \qquad \text{when } \lambda \text{ is expressed in Å and } V \text{ in kilovolts.}$$

The intensity of x radiation rises sharply from zero at this wavelength, increases to a broad maximum, and falls off gradually, as shown in Fig. 1i. But since the reflections are normally recorded on photographic film, the absorption edge of silver ($\lambda = 0.486$ Å) and that of bromine ($\lambda = 0.920$ Å) in the AgBr of the film's emulsion affect the recorded intensity, so that the variation of the effective intensity of the x radiation as used in the Laue method is as shown in Fig. 1ii.

Apparatus. The Laue method employs the simplest instrumentation of all the x-ray methods, basically because the crystal remains in a fixed orientation during the experiment. The essential features of the apparatus are illustrated in Fig. 2. In common with most x-ray methods, a narrow pencil of x radiation is allowed to reach the crystal from the x-ray tube by restricting the radiation with the aid of a *collimator*. The essential parts of this member are two lead plugs mounted in a tube, each containing a small circular aperture called a *pinhole*. The apertures are aligned so as to point at the crystal, which is usually mounted in a device that allows its orientation to be adjusted. The x-ray beam delineated by the collimator is diffracted by the crystal to produce a collection of diffracted beams which are recorded on a photographic film. This is commonly a flat film placed normal to the primary x-ray beam. The direct beam, after continuing on through the crystal, is usually prevented from reaching the photographic film by a lead stop, but this is designed so that it can be removed momentarily to record the location where the direct beam intersects the film.

A photograph of an actual Laue apparatus (commonly called a Laue *camera*) is shown in Fig. 3. The holder for the photographic film, known as a *cassette*, can be removed from the apparatus and taken to the darkroom for development of the film without disturbing the adjustment of the rest of the apparatus with respect to the x-ray source.

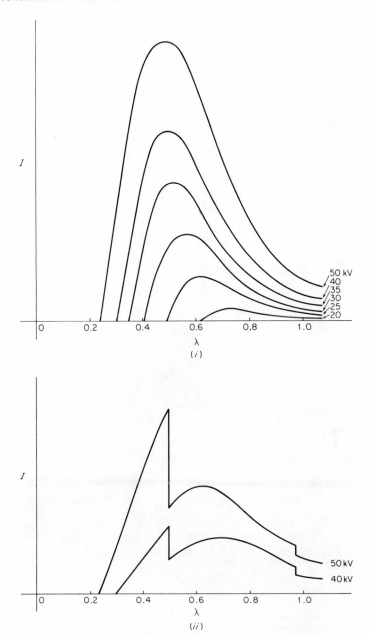

(i)

(ii)

Fig. 1i. The variation of intensity of a source of general x radiation for various potentials across the x-ray tube. [From C. T. Ulrey, *Phys. Rev.* **11** (1918) 407.]

Fig. 1ii. The blackening of a photographic film by x rays produced by potentials of 40 and 50 kV, as a function of wavelength. [From Wyckoff[3], p. 146, modified.]

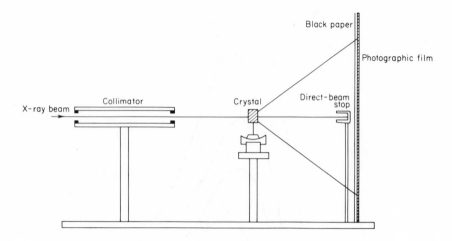

Fig. 2. Diagram of an instrument for making Laue photographs.

Fig. 3. Photograph of an instrument for making ordinary Laue photographs.

Reflection characteristics

One of the characteristics of Laue photographs (which seriously limits their usefulness in x-ray crystallography) is that many of the Laue spots are the superposed records of several orders of reflections. This can be easily demonstrated by examining Bragg's equation in a somewhat different form. Bragg's equation is usually written

$$n\lambda = 2d_{hkl} \sin \theta. \tag{2}$$

In this form the integers h, k, l are the indices of a plane (hkl) which contain no common factor, and n is the order of reflection from this plane. If this same relation is expressed so that the indices are those of a reflection instead of those of a plane, then

The first-order reflection from plane (hkl) is written hkl,
the second-order reflection from plane (hkl) is written $2h\ 2k\ 2l$,
the third-order reflection from plane (hkl) is written $3h\ 3k\ 3l$,
the nth-order reflection from plane (hkl) is written $nh\ nk\ nl$.

When using reflection indices, therefore, (2) can be expressed as

$$\lambda = 2d_{nh\ nk\ nl} \sin \theta, \tag{3}$$

so that the glancing angle is given by

$$\theta = \sin^{-1}\left(\frac{\lambda}{2d_{nh\ nk\ nl}}\right). \tag{4}$$

For a specific wavelength λ_1 and for $n = 1$, this becomes

$$\theta = \sin^{-1}\left(\frac{\lambda_1}{2d_{hkl}}\right). \tag{5}$$

If for λ in (4) there is substituted $\lambda_2 = \frac{1}{2}\lambda_1$, and at the same time a common factor $n = 2$ is inserted in the indices, the result is

$$\theta = \sin^{-1}\left(\frac{\lambda_2}{2d_{2h\ 2k\ 2l}}\right)$$

$$= \sin^{-1}\left(\frac{\frac{1}{2}\lambda_1}{2(\frac{1}{2}d_{hkl})}\right). \tag{6}$$

It is seen that the right of (6) has the same value as the right of (5); this means that the direction of a reflection hkl for wavelength λ is the same as the direction of the reflection $2h\ 2k\ 2l$ with wavelength $\lambda/2$, and more

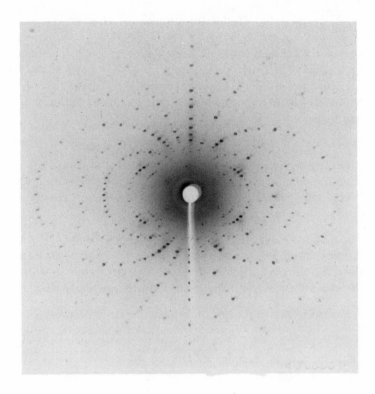

Fig. 4i

Fig. 4. Laue photograph illustrating the occurrence of spots distributed along elliptical loci. The spots along each ellipse are reflections from planes in one zone.

generally, it is the same as the direction of the reflection $nh \; nk \; nl$ for wavelength λ/n. This can be succinctly expressed by saying that θ is invariant for equal values of λ/n. The significance of this invariance is that if the x-ray beam has a sufficient range of wavelengths, any stack of planes (hkl) has the ability to reflect in several orders, all in the same direction defined by θ, because each spacing selects the wavelength appropriate to satisfying the Bragg relation. It is apparent that a Laue spot contains the first-order reflection produced by the wavelength λ_1 from the plane (hkl), plus as many higher orders as there exist submultiples of the λ_1 in the wavelength range used.

It might be thought that this circumstance renders a Laue photograph useless for x-ray crystallography. Although it does produce serious limitations, fortunately the wavelengths of the radiation in the neighborhood of the short-wavelength limit cannot be accompanied by submultiple wavelengths since these would fall beyond the limiting wavelength. Accordingly,

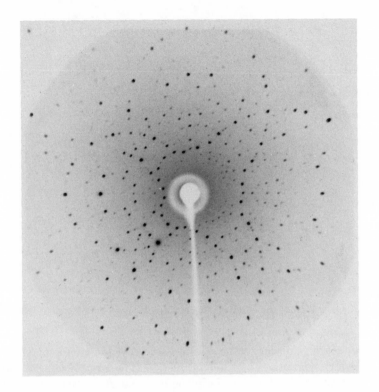

Fig. 4ii

Laue spots produced by wavelengths near this limit contain only first-order contributions. If a small value is substituted for λ in (3), it is evident that a small value of θ is determined, so that these single-component Laue spots lie nearest the center of the Laue photograph. This aspect of Laue spots will be treated in another way later.

General geometry

Zone relations. One of the striking features of most Laue photographs is that the spots appear to be localized along nearly circular curves running through the center of the photograph. This can be seen in Fig. 4. A careful examination shows that these curves are actually ellipses. The spots occurring on the same ellipse prove to be the reflections of planes that lie in the same zone, that is, which contain a common rational crystallographic direction. This can be demonstrated as follows.

In Fig. 5, let some rational direction ST' of the crystal make some small arbitrary angle ψ with the x-ray beam SO, and let plane $UV'W'$ be normal

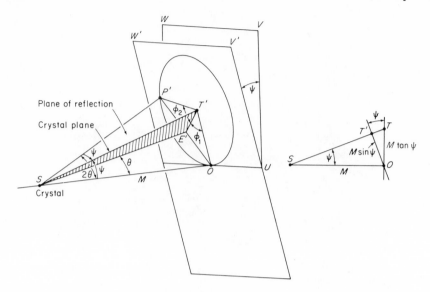

Fig. 5. Derivation of the elliptical shape of the locus of the reflections from planes in one zone.

to this direction. Consider some crystal plane (whose extension is shaded in Fig. 5) which contains the rational direction. The x-ray beam SO and the reflected beam SP' determine the plane of θ and 2θ, namely, plane OSP', which is normal to the crystal plane. The prism $T'SOP'$ is obviously symmetrical in the crystal plane, so that $\phi_2 = \phi_1$ and $T'P' = T'O = M \sin \psi$. Accordingly, P' is always at a distance $M \sin \psi$ from T', so that the locus of all points P' (where the rays reflected by any plane in the zone ST' strike the plane $UV'W'$) is a circle. But the Laue photograph is actually recorded on a plane UVW that is perpendicular to SO and consequently makes an angle ψ with plane $UV'W'$; therefore the circular locus of P' on plane $UV'W'$ projects as an elliptical locus on the plane of the photograph UVW. The semiminor axis of the ellipse is the radius $M \sin \psi$ of the circle, whereas the semimajor axis (which is always along a line radiating from O) is $(M \sin \psi)/\cos \psi = M \tan \psi$.

The locus of rays SP' is a circular cone. As long as $\psi < 45°$, the intersection of this cone with the plane of the photographic film is an ellipse; when $\psi = 45°$ it is a parabola; and when $\psi > 45°$ it is a hyperbola.

Every rational crystal plane contains many rational axes. It follows that every Laue spot is located at the intersection of several conic sections, but most of these may not be very apparent since the ellipses attracting the attention are those with the most closely spaced spots.

Symmetry. When the crystal is oriented so that the x-ray beam is directed along a symmetry element of the Friedel class of the crystal, the Laue photograph displays the projection on the film of the symmetry along this direction. This must be one of the 10 two-dimensional point groups. Determination of the Friedel symmetry is one of the chief uses of Laue photographs. This application is discussed in Chapter 6.

Misorientation with respect to an axis. If the photograph appears nearly symmetrical yet shows lack of complete symmetry of spot positions, especially in the sizes of zone ellipses that ought to be symmetrical, the symmetry element does not contain the x-ray beam, so the crystal is not correctly oriented. The amount of orientation error can be corrected from measurements made on the Laue photographs. An important relation here is that the outer limits of symmetrical ellipses ought to be at equal distances $M \tan \psi$ but actually are found at different distances, say $M \tan \psi_1$ and

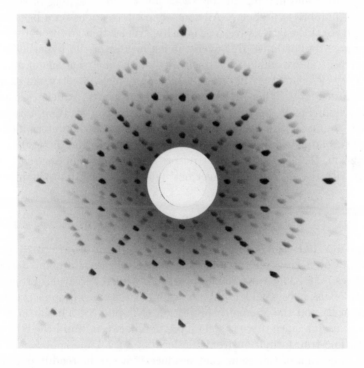

Fig. 6. Back-reflection Laue photograph of a spinel crystal oriented with a four-fold axis along the x-ray beam. Note the four three-fold locations about $3\frac{1}{2}$ cm from the center along diagonals of the square film. Crystal-to-film distance: 3 cm.

$M \tan \psi_2$. The orientation of the crystal must be adjusted in the plane of angles ψ_1 and ψ_2 until these distances are equal. The orientation of crystals is a second important use of the Laue method.

Back-reflection Laue photographs. The customary placement of the photographic film for the Laue method is normal to the x-ray beam and on the opposite side of the crystal from the collimator. An alternative placement that is useful in certain procedures is to place the film normal to the x-ray beam but on the same side as the collimator. This requires passing the collimator through an appropriate light-tight hole in the cassette and its contained film.[5-8,12,28,29]

There are two main reasons for using back-reflection Laue photographs. One is to get a photograph of a crystal that is opaque to x rays or is embedded in an opaque matrix. The other is to determine and correct an orientation error. In the foregoing section on zone relations it was pointed out that for zones making an angle $\psi > 45°$ the intersections of the cones with the film are hyperbolas. All zone-cone intersections on a back-reflection film are of the latter variety. Those hyperbolas that go through the film center are straight lines. But even the curvatures of the hyperbolas that are not central are sufficiently small so that the intersections of zones at symmetry elements are still recognizable as having symmetry. An example is shown in Fig. 6. When the orientation of the crystal is incorrect or unknown, as in Fig. 7, and the axis to be adjusted is a rotation axis in the Friedel symmetry, the amount of misorientation can be determined and the crystal adjusted so that this axis is brought into the direction of the x-ray beam. In this procedure the misorientation can be measured, for example, with the aid of a Greninger net[7] (shown in Fig. 19 of Chapter 9).

Interpretation of reflections

Relation between reflection record and gnomonic projection. Each spot on a Laue photograph corresponds to a crystal plane described by indices (hkl). The same plane can be represented on various projections in common use in crystallography, for example, the stereographic projection (Chapter 2) and the gnomonic projection (Chapter 3). The latter is particularly useful, since there is a close relation between it and the reciprocal lattice. The transformation from the location of a Laue spot to the gnomonic projection of the plane that produces[2,3] it can be readily derived as follows.

Figure 8 is drawn in the plane of angle 2θ and without reference to the orientation of the crystal in space. In the plane of the drawing there occur several important geometrical features, specifically the x-ray beam, the reflection ray, and the normal to the plane (*hkl*). The directions of these three lines are related. In the common version of the Laue method, the photographic film is placed normal to the x-ray beam, so that its position is that of the vertical line in Fig. 8. If the distance from the crystal to the film is taken as unity, then the distance r of the reflected beam from the direct beam, as measured on the film, is

$$r = \tan 2\theta \qquad (7)$$

For convenience in utilizing this later, it can also be written

$$2\theta = \tan^{-1} r. \qquad (8)$$

The normal P to the reflecting plane reaches the vertical line in Fig. 8 at a

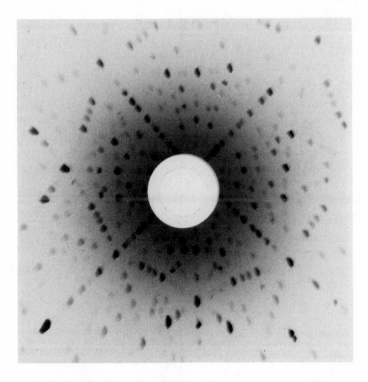

Fig. 7. Back-reflection Laue photograph of the same crystal as in Fig. 6 misaligned by about 8°.

Table 1

Data for scale for transforming Laue spots to gnomonic-projection points
(for crystal-to-film distance, M = 3 cm)

Readings, both scales	Distance to left of pivot	Distance to right of pivot	Readings, both scales	Distance to left of pivot	Distance to right of pivot
5	5 mm	362.3 mm	43	43 mm	57.5 mm
6	6	303.2	44	44	56.8
7	7	261.3	45	45	56.1
8	8	228.9	46	46	55.4
9	9	204.4	47	47	54.7
10	10	184.5	48	48	54.1
11	11	169.0	49	49	53.5
12	12	155.5	50	50	53.0
13	13	144.9	51	51	52.6
14	14	135.1	52	52	52.0
15	15	127.1	53	53	51.5
16	16	120.0	54	54	51.0
17	17	113.7	55	55	50.5
18	18	108.3	56	56	50.1
19	19	103.5	57	57	49.7
20	20	99.1	58	58	49.3
21	21	95.2	59	59	48.9
22	22	91.6	60	60	48.4
23	23	88.4	61	61	48.2
24	24	85.5	62	62	47.8
25	25	82.8	63	63	47.5
26	26	80.4	64	64	47.2
27	27	78.1	65	65	46.9
28	28	76.1	66	66	46.6
29	29	74.2	67	67	46.3
30	30	72.4	68	68	46.0
31	31	71.4	69	69	45.7
32	32	69.2	70	70	45.5
33	33	67.8	71	71	45.2
34	34	66.5	72	72	44.9
35	35	65.2	73	73	44.7
36	36	64.0	74	74	44.5
37	37	62.9	75	75	44.3
38	38	61.9	76	76	44.1
39	39	60.9	77	77	43.8
40	40	59.9	78	78	43.6
41	41	59.1	79	79	43.4
42	42	58.3	80	80	43.2

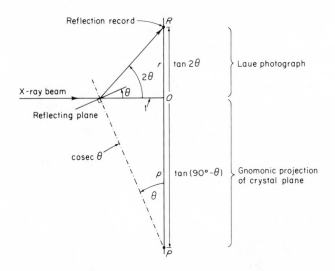

Fig. 8. Relation of the position of a Laue spot to the corresponding gnomonic-projection point.

distance p from the direct beam given by

$$p = \tan(90° - \theta) = \cot \theta. \tag{9}$$

The distances p and r are inversely related; that is, when one is large, the other is small. It will be seen later that it is useful to determine p as a function of r, which is readily done by combining (8) and (9) to obtain

$$p = \cot \tfrac{1}{2} \tan^{-1} r. \tag{10}$$

This relation is based upon a unit crystal-to-film distance. It is often convenient to place the film at another distance, M, instead of unity, and yet measure p in a plane at a unit distance from the crystal. In this event (10) is replaced by

$$p = \cot \tfrac{1}{2} \tan^{-1} (r/M). \tag{11}$$

The simple relation just derived serves to connect the locations of the reflections of a crystal as recorded on a Laue photograph with the locations of the planes of that crystal as they are represented on the gnomonic projection, which is discussed in Chapter 3. With the aid of (11) it is a simple matter to transform the location of each spot that appears at a distance r from the center of the Laue photograph to the gnomonic projection of the plane whose reflection produces it. An example of this kind of transformation is shown in Fig. 9. To make the transformation easy, (11) can be

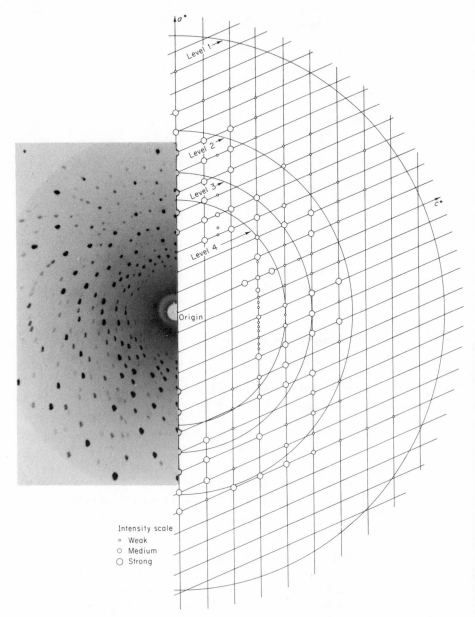

Fig. 9. Laue photograph of petalite, LiAl(Si₂O₅)₂, taken with the x-ray beam parallel to the unique monoclinic axis, and the points of the corresponding gnomonic projection.

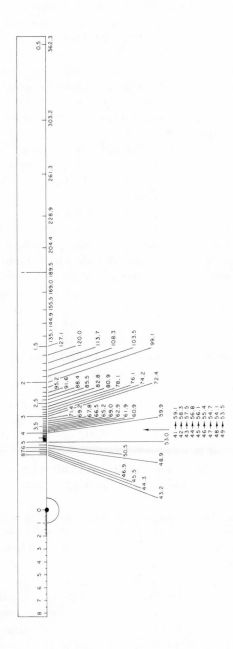

Fig. 10. Scale (modeled after Wyckoff's[2,3]) for transforming the spots of a Laue photograph to the gnomonic projection.

represented graphically by plotting p as a function of r. Another device is to construct a scale (Fig. 10) on one side of which r is marked off as a linear scale, while on the other, corresponding distances p are marked off, each with the label of the corresponding r.

As was pointed out in Chapter 3, the gnomonic projection is related to the reciprocal lattice, for in the gnomonic projection the point representing a plane lies on the normal to the plane, and in the reciprocal lattice the point corresponding to a plane occurs on the same normal. In the first case the distance from the crystal's center is dependent only on the orientation of the plane, and is equal to cosec θ, Fig. 8. In the reciprocal lattice the distance is $1/d_{hkl}$, and the several reciprocal-lattice points corresponding to the various orders of reflection from the same stack of planes constitute a row of points equally spaced along this perpendicular line. Thus, to each point on the gnomonic projection there correspond several reciprocal-lattice points; more specifically, if the several reciprocal-lattice points located on any radial line from the origin are projected along this line and onto the plane of the gnomonic projection, they coincide with a point on the gnomonic projection. In other words, a point on the gnomonic projection of the internal planes of a crystal corresponds to the set of all the reciprocal-lattice points on the radial line to that gnomonic-projection point. The points of the set are hkl, $2h$ $2k$ $2l$, $3h$ $3k$ $3l$ . . . nh nk nl, and these all correspond to the gnomonic-projection pole for plane (hkl). It is apparent that the gnomonic projection contains no inherent record of any common factor in its indices.

In reflection language, the gnomonic projection contains no record of the order of a reflection and may correspond to several orders at once; accordingly, the Laue spot from which it was transformed may contain the superposed contributions of several orders of reflections. This is the same conclusion that was reached in the section "Reflection characteristics". How the gnomonic projection can be used to interpret Laue photographs in spite of this ambiguity will be discussed in the following sections.

Axial photographs. In the foregoing general discussion of the relation of a point on the gnomonic projection to points on the reciprocal lattice, no special orientation of the crystal was assumed. If the orientation of the crystal with respect to the x-ray beam is arbitrary, the interpretation of the gnomonic projection is not straightforward.

On the other hand, if the crystal is oriented so that a rational axis is parallel to the x-ray beam, the relation is a simple one. A basic geometrical relation between the crystal lattice and its reciprocal lattice is that the plane $(uvw)^*$ in the reciprocal lattice is normal to the translation $[uvw]$

in the crystal lattice. Suppose that the crystal axis [001] is oriented parallel to the x-ray beam; then the reciprocal-lattice planes $(00l)$* are normal to the x-ray beam. In these circumstances, the x-ray beam is perpendicular to planes of the reciprocal lattice and to the plane of the gnomonic projection. The parallelism of these planes is responsible for easy interpretation of the gnomonic projection.

Using a construction involving the reciprocal lattice when the wavelength is a variable entails certain complications not encountered with a constant wavelength. If the Ewald construction is used, the reciprocal lattice is plotted so that the origin distances of the points are $\sigma_{hkl} = 1/d_{hkl}$, while the sphere has a variable radius $1/\lambda$ since λ is a variable. The occurrence of a reflection corresponding to a reciprocal-lattice point, say P_3 in Fig. 11, requires a sphere whose surface contains that point. The sphere so determined has a radius $1/\lambda_3$ and has a unique center S_3. In this way the specific value of λ for this reflection is determined. The diffracted ray has the direction S_3P_3. If a reflection arises due to another point, say P_2, this requires still another sphere whose center is at S_2 corresponding to another wavelength λ_2. All possible spheres lie within a limiting sphere of radius $1/\lambda_{\min}$ which is set by the minimum value of λ used in the experiment, as determined by (1).

Whenever two reciprocal-lattice points lie on the same line containing the origin O, such as P_1 and P_2, the similarity of their reflecting geometries

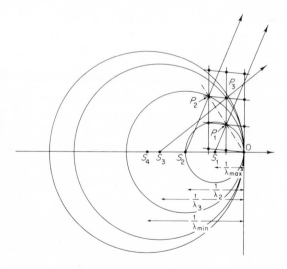

Fig. 11. Relations of Laue reflections to points in the reciprocal lattice.

shows that the diffracted rays have the same direction. This confirms an earlier conclusion that the several orders of reflection from a plane produce the same Laue spot.

The Ewald construction shows that only those reciprocal-lattice points lying within the sphere determined by the minimum wavelength can cause a Laue spot. There is also a somewhat poorly defined practical maximum-wavelength limit set by the bromine absorption edge at $\lambda = 0.920$ Å. The photographic film is not very sensitive to wavelengths greater than this, so that reciprocal-lattice points falling within this sphere may be expected to give rise to weak Laue spots or no spots at all.

While the Ewald construction explains the reflections on Laue photographs, the complete construction becomes cumbersome to use in a routine explanation involving many reflections because of the many circles required and the many sets of multiple parallel rays, such as S_1P_1 and S_2P_2, that contribute to a single spot. A more convenient way to use the Ewald construction is to regard all points within the minimum-wavelength sphere as possible contributors to Laue spots (thus incidentally ignoring the poorly defined maximum-wavelength limit). The reciprocal lattice is then interpreted in terms of the gnomonic projection.

The relation between the properly oriented reciprocal lattice and the gnomonic projection is shown in Fig. 12. In this illustration the view is normal to the x-ray beam, so that the reciprocal-lattice levels are seen edge-on, while the Ewald sphere, drawn with a radius of $1/\lambda_{\min}$, is seen in profile.

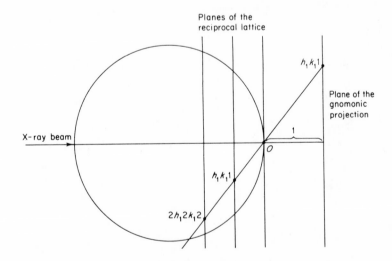

Fig. 12. Relation of Laue spots to levels of the reciprocal lattice.

In general, a line from the origin to some point on the first level, say $h_1k_1 1$, passes through a point on the second level, specifically $2h_1\ 2k_1\ 2$, and in general, through as many points $nh_1\ nk_1\ n$ on each nth level as are contained within the Ewald sphere of radius $1/\lambda_{\min}$. Each reciprocal-lattice point can give rise to a reflection, and all of these reflections are contributed to the same Laue spot, which corresponds to gnomonic point $h_1k_1 1$ (Fig. 12). (No points on the zero level with indices $hk0$ can give rise to a reflection because these points do not fall within the sphere.)

The ambiguity caused by the fact that one gnomonic-projection point often corresponds to several reciprocal-lattice points prevents making a list of observed reflections from data furnished by a Laue photograph, and this limits the use of Laue photographs in space-group determination.

Determination of the reciprocal lattice

Certain regions of a Laue photograph and its corresponding gnomonic projection have spots with first-order components only. There are two categories of such spots.

Laue spots near reflections having simple indices. Reciprocal-lattice points having simple indices are on rays radiating from the origin which have short intervals between points, as shown in Fig. 13. The reflection due to all of these reciprocal-lattice points within the limiting sphere contribute to the same Laue spot, which therefore is usually intense. Because of its simple indices such a spot is at the intersection of several zone ellipses also having simple indices, and which are also usually prominent. But the next neighboring spot positions on the ellipses are found to be locations of missing or very weak spots, as can be seen in Fig. 14. The reason for this, first pointed out by Jeffery[22], is that because of their complicated indices, these neighboring points are on rays with such large intervals between points that none, or only one point per ray, falls within the sphere of minimum wavelength. These points can therefore give rise to spots representing first-order reflections only.

Laue spots with limited orders[11]. The most extensive region having only first-order components is the outer edge of the gnomonic projections between the radii r_1 and r_2 of Fig. 15i. Within this region the points can only correspond with reciprocal-lattice points on the outer edge of the first level, beyond which there are no further points within the minimum-wavelength sphere. These points occur within a hollow cone limited by rays OE_1 and OE_2. All points within this region have indices $hk1$.

Similarly, within the annulus limited by radii r_1 and r_3 only points with indices $hk1$ and $hk2$ occur, and so on. These ranges can be computed from r_1, which corresponds to the outer limit of Laue spots in Fig. 9. This limit cannot be located very accurately, but from it an approximate value of the angle ψ_1 can be obtained; in Fig. 15i

$$\tan \psi_1 = r_1 \tag{12}$$

and, more generally,

$$\tan \psi_n = r_n. \tag{13}$$

These angles can be evaluated from the isosceles triangles SOE, shown in detail in Fig. 15ii.

$$R_n{}^2 = \frac{1}{\lambda^2} - \left(\frac{1}{\lambda} - \zeta_n\right)^2 \tag{14}$$

$$= \zeta_n\left(\frac{2}{\lambda} - \zeta_n\right) \tag{15}$$

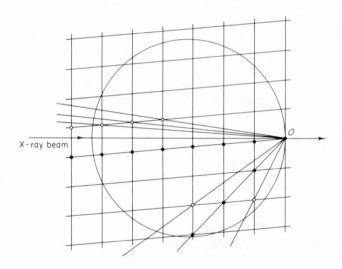

Fig. 13. Geometry used in explaining the blank areas in a Laue photograph that occur in the immediate vicinity of very intense spots with simple indices. ● Reciprocal-lattice point having simple indices; ○ reciprocal-lattice point having complicated indices.

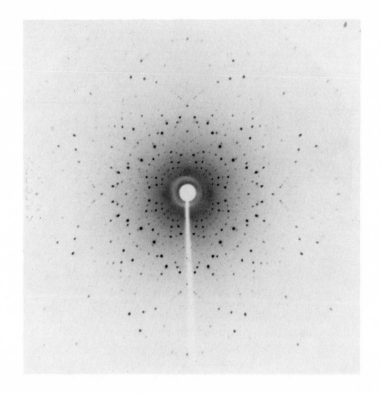

Fig. 14. Laue photograph illustrating consequences of the geometry shown in Fig. 13. Pollucite, a cubic crystal whose composition is approximately $Cs_3NaAl_4Si_8O_{24} \cdot H_2O$.

and

$$\tan \psi_n = \frac{R_n}{\zeta_n} \tag{16}$$

$$\tan^2 \psi_n = \frac{2}{\zeta_n \lambda} - 1 = \frac{2}{n\zeta_1 \lambda} - 1. \tag{17}$$

If (13) and (17) are compared, ψ can be eliminated to give

$$r_n{}^2 = \frac{2}{n\zeta_1 \lambda} - 1. \tag{18}$$

For $n = 1$ this can be solved for the following value of ζ_1.

$$\zeta_1 = \frac{2}{\lambda(r_1{}^2 + 1)}. \tag{19}$$

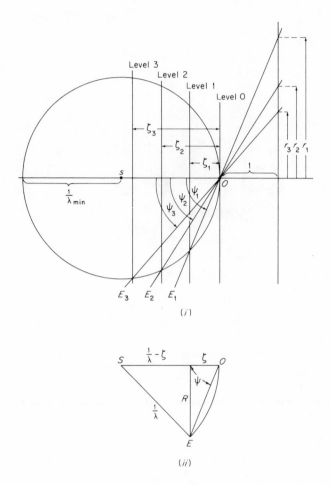

Fig. 15.   Regions of Laue spots with limited orders of reflection.

The last result will prove of importance in a later section. It is dependent upon a measurement of r_1, the radius of the outer limit of the Laue spots on the gnomonic projection. When the value of the resulting ζ_1 is inserted in (18) for $n = 2$, it gives the inner radius of the annulus on the gnomonic projection corresponding to the region on the Laue photograph within which the reflections from only the first level occur. If the x-ray beam is directed along the c axis, all such spots have indices $hk1$. The net determined by these points permits indexing all points on the gnomonic projection of the Laue spots. This was demonstrated in Fig. 9.

When the net determined by the assured $hk1$ reflections is drawn, some points nearer the center of the gnomonic projection are found to have fractional coordinates, such as $1\frac{1}{2}$ $2\frac{1}{2}$ 1. These are points from the upper levels of the reciprocal lattice that have been indexed as if they occurred on the first level. In these instances each of the three indeces must be multiplied by the same number that will transform them each into whole numbers without a common factor. For example, multiplying by 2 produces the transformation $1\frac{1}{2}$ $2\frac{1}{2}$ $1 \rightarrow 3\ 5\ 2$. This is the index of the nearest reciprocal-lattice point to the origin that contributed to the Laue spot corresponding to the gnomonic point $1\frac{1}{2}$ $2\frac{1}{2}$ 1.

Crystal cell. Figure 12 showed that when the crystal is properly oriented, mapping on the first level of the reciprocal lattice is geometrically similar to the mapping on the gnomonic projection. Any feature on the first level occurs mapped on the gnomonic projection with an enlargement ratio of gnomonic/reciprocal lattice $= 1/\zeta_1$. Thus, the net on the gnomonic projection, determined as described in the last section, outlines the reciprocal-lattice net on the first level with scale factor $1/\zeta_1$. Accordingly, the point locations on the first level as well as the origin point of the reciprocal lattice are known. From this information the cell of the reciprocal lattice can be chosen. The dimensions of this cell can be transformed into the dimensions of the crystal cell with the aid of the relations in Table 2 of Chapter 9. Unfortunately, the linear dimensions of the reciprocal cell depend upon ζ_1, and ζ_1 depends on r_1, which cannot be determined accurately. The dimensions of cells determined by the Laue method are accordingly rough values.

Application to the study of thermal motion

Laue photographs often show not only the conventional sharp spots but also some fainter diffuse spots, whose shape and appearance are quite different from those of ordinary Laue spots. These fainter spots are usually in the form of short streaks, sometimes containing an ordinary sharp spot near the center. Preston[13] first showed that these diffuse streaks are related to the temperature motions of the atoms of the crystal structure, for he discovered that the intensities of the diffuse spots increase with temperature.

The explanation of these diffuse spots can be given in terms of the reciprocal lattice. An infinite, perfect crystal has a reciprocal lattice for constant λ that consists of pure geometrical points. If the perfect periodic structure of the same crystal is somewhat disturbed by thermal motion, or by any other positional disorder of its atoms, the reflecting planes are no longer in perfect periodic stacks with uniform spacing d; as a result, each point hkl in reciprocal space is no longer a pure geometrical point, but rather a small finite region with fuzzy edges. The intersection of this region with the sphere of the characteristic radiation produces a diffuse spot in the Laue photograph.

It is not usual to rotate a crystal while taking a Laue photograph, but if this is done, the diffuse spot does not migrate with the Laue spot, showing that the diffuse spot is produced by the characteristic wavelength. This was beautifully demonstrated in a series of photographs by Jauncey and Baltzer.[14]

Thermal agitation has been studied with the aid of the diffuse reflections of Laue photographs. This is discussed in Chapter 12.

Appraisal of the Laue method

The Laue method is often used because it requires minimal apparatus. It therefore provides an easy means of investigating the Friedel symmetry. In this application, however, it lacks the elegance of permitting separate study of the symmetries of the zero and upper levels, which is possible with the Weissenberg and precession methods.

Space-group symmetry is normally approached through a study of absent reflections, which leads to a knowledge of the diffraction symbol[t§]. When the symmetry operations of a space group contain translations that are submultiples of the cell-edge translations, certain reflections whose indices are based upon this cell are systematically absent. When the cell is non-primitive (i.e., "centered"), systematic absences, dependent on the lattice type, occur among the general hkl reflections. Glide planes cause systematic absences in reflections like $hk0$, $h0l$, $0kl$, and $hh0$, which permit the presence and the locations of glide planes to be ascertained. Screw axes cause absences in reflection like $h00$, $0k0$, $00l$, and $hh0$, which permit the presence and locations of the screw axes to be ascertained.

Laue photographs are normally taken with the direct x-ray beam along a crystallographic axis. In such circumstances only reflections from planes $(hk1)$, $(h1l)$, and $(1kl)$ are recorded. Among these only a limited number

---

[t] M. J. Buerger. X-ray crystallography (Wiley, New York, 1942) 510–516.

[§] M. J. Buerger. Diffraction symbols. Chapter 3 of "Physics of the solid state" edited by S. Balakrishna, B. Ramachandra Rao, and M. Krishnamurthy (Academic Press, London, 1969) 27–42.

within a small ring near the center of the photograph are present in the first order only. Ordinarily, these are in sufficient numbers so that absences among them can be used to fix the lattice type. But the glide planes and screw axes, which depend on zero-level reflections, cannot be fixed by data from such photographs. Even if additional photographs are taken with other directions of the x-ray beam, the presence of screw axes can never be fixed, for the absences characteristic of them occur in rows $[h00]^*$, $[0k0]^*$, and $[00l]^*$, which are densely packed with reciprocal-lattice points and which are therefore never available in first-order reflections only. Also, if special directions of the x-ray beam are used to reveal some of the more general reflections on the zero levels, a sufficient number of them cannot be recorded in one order to determine the presence or absence of glide planes. Thus, the determination of diffraction symmetry from Laue photographs is limited to the determination of the Friedel symmetry and the lattice type.

The dimensions of the unit cell can be established roughly from Laue photographs. The reason they cannot be established exactly is fundamentally because a linear dimension depends on the magnitude of d in Bragg's equation, which is a function of the particular λ causing the reflection, and this is unknown. The rough value can be determined from a knowledge of the minimum wavelength corresponding to the potential across the x-ray tube, which is not very accurately known, or from the radius within which all gnomonic poles lie, which is also not accurately determinable. The axial ratio of the cell, however, is accurately determinable.

It is practically impossible to use the Laue method in locating the atoms in the cell because the theory for accomplishing this requires an accurate knowledge of the fraction of the original x-ray intensity recorded in each reflection. But in the Laue method the original x-ray beam varies in an unknown way according to the wavelength selected by the particular d_{hkl}. Furthermore, the intensities of individual reflections hkl are required, whereas in the Laue method several orders record together in the same spot. Because of these difficulties, the Laue method is not used for crystal-structure analysis.

The chief application of the Laue method in the past has been in the investigation of the Friedel symmetry and in the determination of the orientation of a crystal. Metallurgists commonly use it in routine testing. A more sophisticated and very fruitful use has been in recording diffuse scattering in connection with studying the thermal motion of molecules in organic crystals.

A hitherto unrecognized application of the Laue method is in the identification of unknown crystals, commonly called crystallochemical analysis. This is discussed in Chapter 7.

Notes on history

The first x-ray diffraction experiment tried by Friedrich and Knipping[1] was an attempt to make a back-reflection photograph with a copper sulfate crystal. No diffraction effect was observed, so this first experiment was a failure. In the second experiment photographic plates were placed in the front-, side-, and back-reflection regions. The front-reflection plate showed the essential features of what is now regarded as a standard Laue photograph.

Laue photographs were used in the early days of x-ray crystallography to establish the Friedel symmetry of the crystal and to determine the approximate dimensions of the cell. The latter use is now rare since much more accurate cell dimensions are easily determined with the aid of rotation, Weissenberg, or precession photographs.

Murdock[11] was the first to point out how Laue photographs can be indexed with the aid of the reciprocal lattice. Greninger[7,8] first called attention to the advantages of the back-reflection Laue photograph in determining and adjusting the orientation of a crystal.

Significant literature

1. W. Friedrich, P. Knipping, and M. Laue. Interferenz-Erscheinungen bei Röntgenstrahlen. (*Sitzungsberichte der mathematischphysikalischen Klasse der Königlich Bayerischen Akademie der Wissenschaften zu München*, 1912) 303–322.
2. Ralph W. G. Wyckoff. The crystal structures of some carbonates of the calcite group. *Am. Jour. Sci.* **50** (1920) 317–360, especially 318–332.
3. Ralph W. G. Wyckoff. The structure of crystals. (Chemical Catalog Co., New York, second edition, 1931) 124–150.
4. E. Schiebold. Die Lauemethode. Vol. I of "Methoden der Kristallstrukturbestimmung mit Röntgenstrahlen" (Akademische Verlagsgesellschaft, Leipzig, 1932) 173 pages.
5. W. Boas and E. Schmid. Laue-Diagramme mit grossen Ablenkungswinkeln. *Metallwirtschaft* [now "*Metall*"] **10** (1931) 917–919.
6. Ludwik Chrobak. Die "zurückreflektierten" Laue-Interferenzbilder. *Z. Kristallogr.* **82** (1932) 342–347.
7. Alden B. Greninger. Determination of the orientation of metallic crystals by means of back-reflection Laue photographs. *Trans. Am. Inst. Min. and Met. Eng.* **117** (1935) 61–71.
8. Alden B. Greninger. A back-reflection Laue method for determining crystal orientation. *Z. Kristallogr.* **91** (1935) 424–432.
9. G. L. Clark and S. T. Gross. A new type of gnomonic ruler. *Science* **86** (1937) 272–273.
10. Franz Halla and Hermann Mark. Leitfaden für die Röntgenographische untersuchung von Kristallen. (Johan Ambrosius Barth, Leipzig, 1937) 193–208.
11. Carleton C. Murdock. The interpretation of Laue photographs in terms of the reciprocal lattice. *Z. Kristallogr.* **99** (1938) 205–216.

12. D. E. Thomas. Laue patterns by reflected x-rays. *Journ. Sci. Instrum.* **16** (1939) 222–228.
13. G. D. Preston. Diffraction of x-rays by crystals at elevated temperatures. *Proc. Roy. Soc. London* **A172** (1939) 116–126.
14. G. E. M. Jauncey and O. J. Baltzer. Non-Laue maxima in the diffraction of x-rays from rocksalt—equatorial maxima. *Phys. Rev.* **59** (1941) 699–705.
15. Beulah Field Decker. A simple vector method for the determination of orientation of cubic single crystals from back-reflection x-ray photographs. *J. Appl. Phys.* **15** (1944) 610–612.
16. Samuel G. Gordon. Simple gnomonic projector for x-ray lauegrams. *Amer. Mineralogist* **33** (1948) 634–638.
17. Tentative method for determining the orientation of a metal crystal. *Amer. Soc. Test. Mat.* **E82–49T** (1949) 1105–1117.
18. N. F. M. Henry, H. Lipson, and W. A. Wooster. The interpretation of x-ray diffraction photographs. (Macmillan, London, 1951) 71–86.
19. William H. Barnes and Sydney Wagner. An x-ray diffraction Laue microcamera. *Canadian Jour. Technology* **29** (1951) 337–342.
20. B. D. Cullity. Elements of x-ray diffraction. (Addison Wesley, Reading, Massachusetts, 1956) 89–92, 138–148, 215–236, 502–505.
21. D. Griffiths and A. Franks. Laue rotation lines. *Acta Cryst.* **11** (1958) 53.
22. J. W. Jeffery. An investigation of the blank areas on Laue photographs round: 1. the direct beam, and 2. reflections with simple indices. *Z. Kristallogr.* **110** (1958) 321–328.
23. Lorin L. Hawes. A gnomonic projector. *Amer. Mineralog.* **49** (1964) 180–183.
24. M. Oron and I. Minkoff. A novel form of x-ray diffraction microbeam camera. *J. Sci. Instrum.* **42** (1965) 337–338.
25. J. Rioux. Tête goniométrique de grande capacité pour l'orientation (méthode de Laüe) le repérage et le découpage d'un monocristal selon une direction prédéterminée. *Bull. Soc. franç. Minér. Crist.* **89** (1966) 329–332.
26. Masakazu Murakami. One method of visualizing the Laue pattern. *Japan J. Appl. Phys.* **5** (1966) 448.
27. B. H. Matzinger and P. M. de Wolff. An optical instrument for the direct interpretation of Laue patterns. *Acta Cryst.* **22** (1967) 764–765.
28. C. Arguello. New conical camera for single-crystal orientation by means of x-rays. *Rev. Sci. Instrum.* **38** (1967) 598–600.
29. C. F. Sampson and C. Wilkinson. The use of a conical back reflexion camera for single-crystal orientation. *J. Appl. Cryst.* **2** (1969) 138–140.

Chapter 6

Symmetry determination with the aid of Laue photographs

The Laue method is, in essence, a device for studying crystals by the incidence of x rays along a given direction. A consequence is that a Laue photograph shows the symmetry of this direction. This is an important advantage of the method because it allows one to recognize elements of symmetry in the crystal by direct inspection of the photograph and to determine the Friedel point group (discussed in the next section) to which the crystal conforms. In order to be able to accomplish this, more than one Laue photograph must be taken. This cannot be done by taking photographs along directions selected at random. Rather, to establish Friedel symmetry, the incident x-ray beam must be directed along crystal directions of high symmetry. In this chapter the symmetry that can be recognized directly from Laue photographs is discussed.

Friedel symmetry

It is well known that every crystal has a megascopic symmetry which conforms to one of the 32 crystallographic point groups. When the first Laue photographs were made, it came as a surprise[1,2] that the photographs of crystals with symmetries $\bar{4}3m$ and $4/m\,\bar{3}\,2/m$ showed the same symmetry, namely $4/m\,\bar{3}\,2/m$. As early as December, 1913, George Friedel[3] pointed out that not all the 32 symmetries could be determined by the

123

inspection of x-ray photographs of a crystal. The reason for this is basically that reflections hkl and $\bar{h}\bar{k}\bar{l}$ are identical in magnitude although they do differ in phase. Since only the magnitude of a reflection can be measured, x-ray methods cannot distinguish between reflection from the top of a stack of planes indexed as (hkl) and that from the bottom of the same stack. Accordingly, the diffraction effects of a crystal are always the same as if the crystal had a center of symmetry, whether it does have such a center or not. Friedel enunciated the effect in a curious negative way essentially as follows. *An x-ray photograph can never reveal the absence of a center of symmetry.* This precept became known as *Friedel's law*, although it is usually stated in the form that x-ray diffraction effects cannot reveal whether a crystal has or has not a center of symmetry[†].

In any event, x-ray diffraction effects are the same for reflections hkl and $\bar{h}\bar{k}\bar{l}$, so one can only distinguish crystal symmetries as if they were the 11 centrosymmetrical crystallographic point groups. Accordingly, if the symmetry of a crystal does not include a center of symmetry, the crystal gives the same observable diffraction effect as if it had a center. All crystal classes that have the same diffraction symmetry are said to belong to the same *Friedel class*. The way each of the 32 crystallographic symmetries, when augmented by a center of symmetry if one is not already present, becomes one of the 11 Friedel symmetries is outlined in Table 1.

Some interesting details of the transformations of the 32 crystallographic point groups into 11 centrosymmetrical point groups are outlined in Table 2. Point groups can be subdivided into two major categories called *monaxial* and *polyaxial*. In each point group of the monaxial category, there is only one direction that has axial symmetry, whereas in the polyaxial point groups, axial symmetry occurs along several directions. In each of these two categories, the axes can exclude operations of the second sort, or they can include some operations of the second sort. Those classes which exclude such operations can display enantiomorphism[§], while those which include them cannot display enantiomorphism.

Eleven of the classes contain only pure rotation axes, that is, axes of the first sort; these classes, listed in columns 2 and 5 of Table 2, are called enantiomorphic[§]. When a center of symmetry is added to any of these classes, a related point group containing a center of symmetry results. The 11 resulting point groups are listed in columns 4 and 7 of Table 2. If a

---

[†] This holds true only when anomalous dispersion does not occur, or when its effects are negligible.

[§] If an object and its mirror image are not congruent, they are called *enantiomorphs*. The relation between them is called *enantiomorphism*. Each member of a pair of enantiomorphs is said to be *enantiomorphic* to the other.

center of symmetry is added to the nonenantiomorphic, noncentrosymmetrical classes, the symmetries obtained coincide with one of the preceding centrosymmetrical classes. Therefore, the symmetries that can be determined by x-ray diffraction are only those of the 11 centrosymmetrical classes already noted, that is, those of the nonenantiomorphic centrosymmetrical crystal classes, the *Friedel classes*. Accordingly, hemimorphic and enantiomorphic classes cannot be distinguished by x-ray diffraction.

The addition of a center of symmetry to a crystal that has a principal axis of even order introduces a plane of symmetry perpendicular to that axis. Consequently, if twofold axes exist perpendicular to the principal axis, planes of symmetry that contain the principal axis also appear in the Friedel symmetry. Unless this is recognized, some confusion may arise in practice when the symmetry of a crystal is determined with the aid of x-ray diffraction experiments.

The symmetry of the Laue photograph

The Friedel law merely requires the intensity function to have a center of symmetry at the origin of the reciprocal space lattice. The Laue method, imposing a fixed incidence of the x rays on the reciprocal lattice, cannot show the centrosymmetry of the whole diffraction space in a single photograph. The centrosymmetry of diffraction space can be shown, however, by taking two Laue photographs of a crystal, the second obtained after rotating the crystal 180° from the original one. The Laue photographs so obtained are centrosymmetrical equivalents of each other. Accordingly, a Laue photograph of a crystal taken at random orientation with respect to the x-ray beam does not show, in general, the center of symmetry introduced by the experiment. However, the consequences of the center of symmetry introduced into the reciprocal lattice are vital since the total symmetry of all diffraction space is then consistent with one of the 11 Friedel classes.

Conventional Laue photographs taken with a flat photographic film perpendicular to the x-ray beam give direct information about the Friedel symmetry of that direction. This symmetry embraces both the axial symmetry of the direction and any mirror symmetry containing the axis. The entire symmetry of the x-ray-beam direction in the crystal is seen as its projection on the perpendicular photographic film. The only finite symmetries that can be displayed in two dimensions are the 10 crystallographic point groups in a plane, shown in Fig. 1. A conventional Laue photograph, and its normal projection, therefore, are limited to one of these 10 symmetries. Figures 2–11 show these various symmetries as displayed by actual Laue photographs.

Table 1

Transformations of the 32 point groups into the 11 centro-symmetrical point groups when augmented by a center of symmetry

Crystal system	Point groups	Augmenting operation	Resulting centrosymmetrical point group
Triclinic	1 $\bar{1}$	$\bar{1}$	$\bar{1}$
Monoclinic	2 m $\dfrac{2}{m}$	$\bar{1}$ $\bar{1}$	$\dfrac{2}{m}$
Orthorhombic	2 2 2 $m\,m\,2$ $\dfrac{2}{m}\ \dfrac{2}{m}\ \dfrac{2}{m}$	$\bar{1}$ $\bar{1}$	$\dfrac{2}{m}\ \dfrac{2}{m}\ \dfrac{2}{m}$
Tetragonal	4 $\bar{4}$ $\dfrac{4}{m}$	$\bar{1}$ $\bar{1}$	$\dfrac{4}{m}$
	4 2 2 4 $m\,m$ $\bar{4}\,2\,m$ $\dfrac{4}{m}\ \dfrac{2}{m}\ \dfrac{2}{m}$	$\bar{1}$ $\bar{1}$ $\bar{1}$	$\dfrac{4}{m}\ \dfrac{2}{m}\ \dfrac{2}{m}$

Table 1 (continued)

Crystal system	Point groups	Augmenting operation	Resulting centrosymmetrical point group
Hexagonal	3 $\bar{3}$	$\bar{1}$	$\bar{3}$
	32 $3\,m$ $\bar{3}\,\dfrac{2}{m}$	$\bar{1}$ $\bar{1}$	$\bar{3}\,\dfrac{2}{m}$
	6 $\bar{6}$ $\dfrac{6}{m}$	$\bar{1}$ $\bar{1}$	$\dfrac{6}{m}$
	$6\,2\,2$ $6\,m\,m$ $\bar{6}\,2\,m$ $\dfrac{6}{m}\,\dfrac{2}{m}\,\dfrac{2}{m}$	$\bar{1}$ $\bar{1}$ $\bar{1}$	$\dfrac{6}{m}\,\dfrac{2}{m}\,\dfrac{2}{m}$
Isometric	23 $\dfrac{2}{m}\,\bar{3}$	$\bar{1}$	$\dfrac{2}{m}\,\bar{3}$
	$4\,3\,2$ $\bar{4}\,3\,2$ $\dfrac{4}{m}\,\bar{3}\,\dfrac{2}{m}$	$\bar{1}$ $\bar{1}$	$\dfrac{4}{m}\,\bar{3}\,\dfrac{2}{m}$

Table 2

The 32 point groups grouped according to centrosymmetry,
Friedel symmetry, and enantiomorphism

System	Monaxial groups			Polyaxial groups		
	Enantio-morphic	Nonenantiomorphic		Enantio-morphic	Nonenantiomorphic	
Triclinic	1	$\bar{1}$				
Monoclinic	2	$m = \bar{2}$	$\dfrac{2}{m}$			
Orthorhombic				222	$mm2$	$\dfrac{2}{m}\dfrac{2}{m}\dfrac{2}{m}$
Tetragonal	4	$\bar{4}$	$\dfrac{4}{m}$	422	$4mm$ $\bar{4}2m$	$\dfrac{4}{m}\dfrac{2}{m}\dfrac{2}{m}$
Hexagonal	3	$\bar{3}$		32	$3m$	$\bar{3}\dfrac{2}{m}$
	6	$\bar{6}$	$\dfrac{6}{m}$	622	$6mm$ $\bar{6}2m$	$\dfrac{6}{m}\dfrac{2}{m}\dfrac{2}{m}$
Cubic				23		$\dfrac{2}{m}\bar{3}$
				432	$\bar{4}3m$	$\dfrac{4}{m}\bar{3}\dfrac{2}{m}$
	Noncentro-symmetrical	Centro-symmetrical		Noncentro-symmetrical	Centro-symmetrical	

Friedel
symmetries

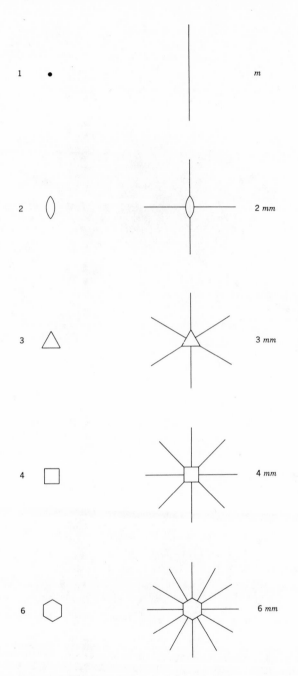

Fig. 1. The 10 crystallographic point groups in a plane. [From M. J. Buerger. Elementary crystallography (Wiley, New York, 1963) 70.]

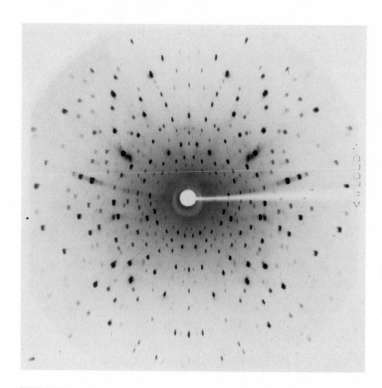

Figs. 2–11. Examples of the 10 two-dimensional symmetries shown by Laue photographs. Natural size; crystal-to-film distance 3 cm.

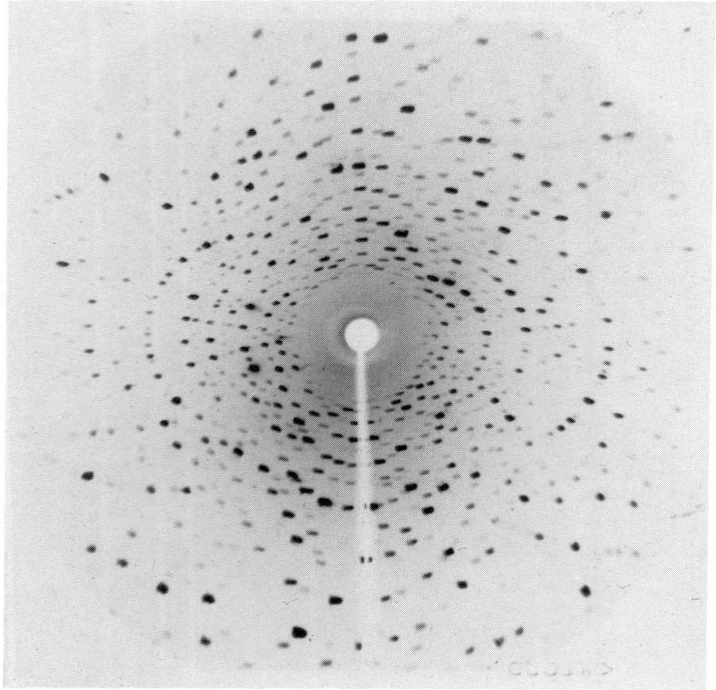

Fig. 3. Symmetry *m*: tetracyano-1, 4 dithiin.

Fig. 2. Symmetry 1: axenite.

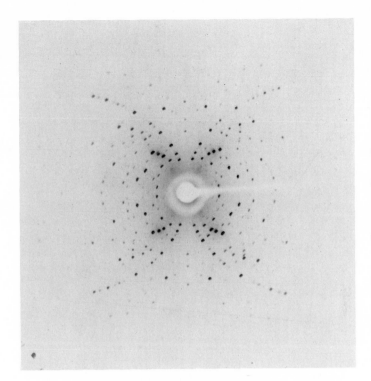

Fig. 5. Symmetry 2*mm*: fluorenon.

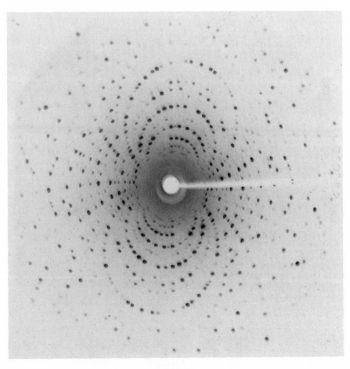

Fig. 4. Symmetry 2: kaliborite.

131

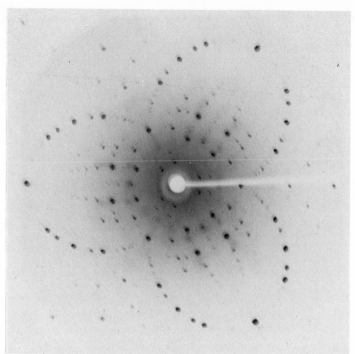

Fig. 7. Symmetry 3*m*: calcite.

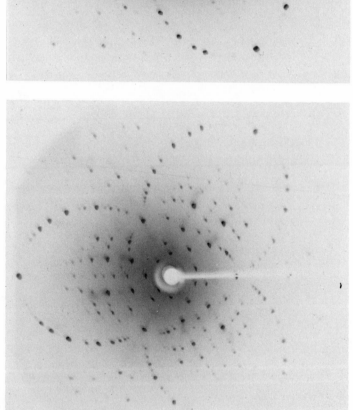

Fig. 6. Symmetry 3: dolomite.

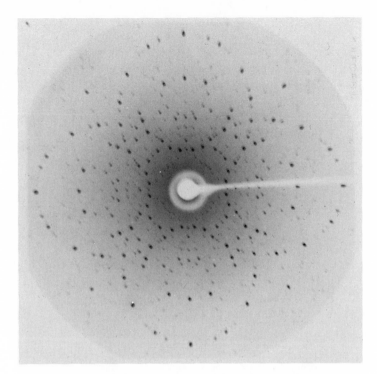

Fig. 9. Symmetry 4*mm*: garnet.

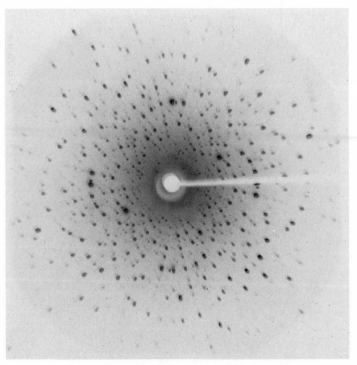

Fig. 8. Symmetry 4: scapolite.

133

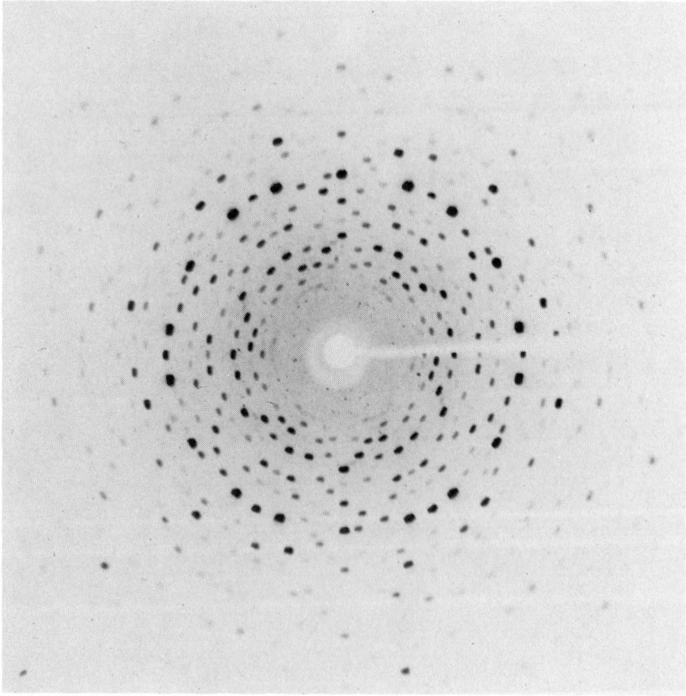

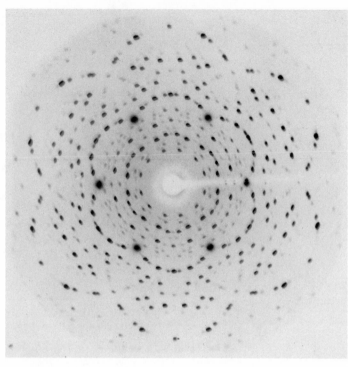

Fig. 11. Symmetry 6*mm*: beryl.

Fig. 10. Symmetry 6: apatite.

Table 3

*The projection symmetries of Laue photographs taken with
the x-ray beam in specified directions*

System	Friedel class	Direction of x-ray beam						
		[100]	[010]	[001]	[0vw]	[u0w]	[uv0]	[uvw]
Triclinic	$\bar{1}$	1	1	1	1	1	1	1
Monoclinic (first setting)	$\dfrac{2}{m}$	m	m	2	1	1	m	1
Ortho-rhombic	$\dfrac{2}{m}\dfrac{2}{m}\dfrac{2}{m}$	2mm	2mm	2mm	m	m	m	1

System	Friedel class	[100]	[001]	[110]	[uv0]	[u0w]	[uuw]	[uvw]
Tetragonal	$\dfrac{4}{m}$	m	4	m	m	1	1	1
	$\dfrac{4}{m}\dfrac{2}{m}\dfrac{2}{m}$	2mm	4mm	2mm	m	m	m	1

System	Friedel class	[10.0]	[00.1]	[11.0]	[uv.0]	[u0.w]	[uu.w]	[uv.w]
Hexagonal	$\bar{3}$	1	3	1	1	1	1	1
	$\bar{3}\dfrac{2}{m}$	m	3m	2	1	m	1	1
	$\dfrac{6}{m}$	m	6	m	m	1	1	1
	$\dfrac{6}{m}\dfrac{2}{m}\dfrac{2}{m}$	2mm	6mm	2mm	m	m	m	1

System	Friedel class	[100]	[111]	[110]	[uv0]	[uuw]	[uvw]
Cubic	$\dfrac{2}{m}\bar{3}$	2mm	3	m	m	1	1
	$\dfrac{4}{m}\bar{3}\dfrac{2}{m}$	4mm	3m	2mm	m	m	1

The actual three-dimensional Friedel symmetry of the crystal must be examined in several projections, each in two dimensions. The ways the 11 specific Friedel symmetries appear in the various important directions of the various crystal systems are displayed in Table 3.

The symmetry of the photograph is usually obvious, so that the Laue method is very useful for determining the Friedel symmetry of the crystal projected parallel to a rational direction. By taking several conventional Laue photographs along appropriate directions, the total Friedel symmetry of the crystal can eventually be determined. In general, a photograph taken with the x-ray beam parallel to a general direction (i.e., not lying on any element of symmetry) has no symmetry. The highest symmetry can be obtained by inspecting directions of lowest indices.

Laue diagrams taken with the x-ray beam parallel to equivalent crystallographic directions are expected to be the same. However, due to Friedel's law, an x-ray diffraction experiment causes Laue photographs to appear identical when taken with the x-ray beam parallel to directions that are equivalent in the Friedel class; nevertheless, these directions need not be truly equivalent in the actual point group of the crystal, because Friedel's law may add new symmetry elements.

In a diagram taken with the beam along the c axis, crystals with 3 and $\bar{3}$ axes can be readily distinguished from crystals with 6 and $6/m$ axes.

The symmetry of the cylindrical Laue photograph

If a cylindrical film with its axis perpendicular to the incident beam is used instead of a flat film, the amount of directly observable symmetry information is reduced, owing to the lower symmetry of the cylinder as compared with the flat plate. In fact, the elements of symmetry that can be directly recognized are mirror planes coinciding either with the plane determined by the camera axis and the x-ray beam or with the plane perpendicular to the camera axis. A twofold axis parallel to the direction of the x-ray beam appears as an inversion center located at the incidence point of the x-ray beam. Other elements of symmetry cannot be readily recognized even if they are parallel to the x-ray beam.

The cylindrical Laue photograph, however, has the advantage that all the information about the crystal is collected at once and on one film; information is obtained not only for the symmetry along the direction parallel to the beam, but also for that along any other direction of the crystal inside the Ewald sphere that corresponds to λ_{min}. To retrieve such information through the study of the crystal projection requires only a proper analysis of that projection. Therefore, in many cases, the cylindrical

Laue arrangement is recommended for the study of the Friedel symmetry of the crystal, for instance, when dealing with small crystals; this arrangement will be discussed at length in Chapter 9. The study of the symmetry of a crystal from a cylindrical Laue photograph requires some knowledge of the basic concepts of geometrical crystallography, for the symmetry can only be recognized through the analysis of the angular relations between poles and zones in the projection. The basic problems encountered in the interpretation of the cylindrical Laue photographs are the same ones that were developed by morphological crystallographers; these are treated in detail in Chapter 9.

Significant literature

1. W. Friedrich, P. Knipping, and M. Laue. Interferenzerscheinungen bei Röntgenstrahlen. *Ann. Phys.* **41** (1913) 971–1002, especially 984–986.
2. M. Laue. Die dreizälig-symmetrischen Röntgenstrahlaufnahmen an regulären Kristallen. *Ann. Phys.* **42** (1913) 397–414.
3. G. Friedel. Sur les symétries cristallines que peut révéler la diffraction des rayons Röntgen. *Compt. rend.* **157** (1913) 1533–1536.
4. M. v. Laue. Über die Symmetrie der Kristall-Röntgenogramme. *Ann. Phys.* **56** (1916) 433–446.
5. W. H. Barnes and A. V. Wendling. Note on the Laue symmetry exhibited by orthogonal crystals. *Amer. Mineralogist* **20** (1935) 253–259.

Chapter 7

Crystallochemical analysis

Background

One of the oldest generalities in crystallography is the law of constancy of interfacial angles for a particular crystalline substance. This was appreciated by Nicolaus Steno in 1669 and made abundantly clear in 1772 after Romé de l'Isle had confirmed it by measurement of many crystals.

This generality implies a corollary, that crystals of different substances are characterized by different interfacial angles. So it was natural that the idea should occur to some crystallographer that a substance in crystalline form ought to be identifiable by its characteristic interfacial angles. Apparently, the first to express this view was Frankenheim[1], as early as 1842.

The first serious attempt to identify crystals from information derived from measuring angles that locate crystal faces was made by the great Russian crystallographer Fedorov[2], who also compiled a monumental systematic catalog, called *Das Krystallreich*[3]; this work incorporated data based upon all crystals that had been measured to 1911. Fedorov devised a routine for identifying crystalline substances for which he coined the name *crystallo-chemical analysis*. This routine was based upon his ideas of crystal structure; unfortunately, it involved tedious computations leading to what he regarded as the best lattice to explain the observed crystal morphology.

Fedorov's identification system was successful in the master's hands but was too tedious and complicated to gain any popularity. One of his students, T. V. Barker, was persuaded to develop Fedorov's method, but came to the conclusion that it was not a practicable routine. Instead, Barker developed a simpler method[7] based on the more obvious geometry of face development of crystals. Unfortunately, Barker died in 1931 before he was able to publish the full manuscript of his method. A group of other crystallographers banded together to test the method, which they found workable, and eventually published several volumes of *The Barker index of crystals*[22] under the editorship of M. W. Porter and R. C. Spiller. The general nature of the method is treated in Volume 1, Part I of that work, and outlined by Donnay[21] in *Physical methods of organic chemistry*. Meanwhile, another of Fedorov's students, A. K. Boldyrev, devised an alternative simplification[5,12,15,18] of his teacher's scheme of identifying crystalline substances.

None of these identification methods came into general use. There are a number of reasons for this. Whereas Fedorov's original method was too complicated, Barker's attempt at a simpler method involved the necessity of learning a curious artificial brand of crystallography and many artificial rules to make it succeed in the identification of crystals. A second reason is that, with the advent of examining crystals by x-ray diffraction methods, and particularly by the powder method, it was inevitable that crystals should come to be identified by their powder patterns. The contrast between this simple method and Barker's complicated, artificial crystallography with its arbitrary rules was so extreme that few except dedicated crystallographers came to know anything about crystallochemical analysis.

The failure of the older methods of crystallochemical analysis is fundamentally attributable to the fact that they rely on the faces that make up the surface morphology as an adequate representation of the rational planes of the lattice of a crystal. The implication here is that, from a knowledge of observable external planes of a crystal, the structural cell of the crystal can be deduced in respect to its axial ratios and lattice type. This implies, in turn, that there is a simple theory that relates the distribution of surface planes to cell and lattice type. Assuming that an adequate theory of this connection exists, it was this very feature that caused the tedious calculations in the Fedorov method. It was also this feature that the Barker method attempted to avoid by an appeal to a rule of simplest indices to determine what amounts to an acceptable cell.

Appraisal of the powder method

Although the identification of crystalline compounds by their x-ray powder-diffraction patterns has extreme simplicity to recommend it, the method has some serious disadvantages. Most importantly, the powder pattern does not, in itself, provide any data except a sequence of values of the interplanar spacings, many of which are often unresolved from one another. No symmetry information is obvious from the appearance of the record, so the one-dimensional list of d_{hkl}'s cannot necessarily be transformed into any useful primary crystallographic variables, such as unit-cell dimensions, except when the sequence of d values corresponds exactly with one of the 17 sequences[†] expected for isometric crystals. Thus, identification by the powder method requires a set of arbitrary, though simple, rules that are without obvious crystallographic significance.

These disadvantages of the powder method arise because it makes use of a sample consisting of an aggregate of crystals, so that the basic geometry of the crystal, which would be obvious if the d_{hkl}'s were obtained from a single crystal, is lost. Whenever it is not necessary to work with an aggregate, that is, whenever a single crystal is available, then this basic geometry is available and ought to be preserved.

Single-crystal x-ray methods of identification

From a philosophical point of view, a crystal structure is completely identified by its Fourier transform, so that identification by means of x-ray diffraction theoretically requires data for constructing the Fourier transform of the crystal structure. This requirement implies a knowledge of the cell dimensions, the space-group symmetry, and the structure amplitudes F_{hkl} in both magnitude and phase. Although none of these data can be supplied directly by the powder method, appropriate single-crystal methods can supply the cell dimensions, diffraction symbol, and the magnitudes of the F_{hkl}'s. If homometric structures are ignored, lack of phase data does not exclude complete correlation with a specific crystal structure. Thus, single-crystal methods are capable of providing all crystallographically significant data, and are potentially useful in crystallochemical analysis. For this purpose, the most powerful single-crystal methods are the Weissenberg and precession methods, which make use of moving films.

With the recent increase of interest in devices that make use of the so-called solid-state properties of crystals, many large industrial research laboratories, as aids in their programs of finding crystals with desirable properties, have been engaged in growing single crystals of many kinds. When possible, these crystals have been examined for symmetry and cell dimensions by the precession method, a procedure leading to a knowledge

[†] J. D. H. Donnay and Gabrielle Donnay, Reflections permitted by each of the 17 cubic aspects. "International tables for x-ray crystallography crystallography," vol. II (1959) 147–149.

of the diffraction symbol of the crystal[†]. In order to identify the crystal, this information has been compared with the Donnay tables,[21] which list known crystalline compounds by systems in order of increasing values of axial ratio or the cube edge. In this way, a sophisticated equivalent of Fedorov's crystallochemical analysis has spontaneously arisen without any general notice.

Because of the exacting mechanical construction required for building moving-film single-crystal apparatus, and the rather general ignorance among chemists of the theory of using such apparatus, the Weissenberg and precession methods of crystallochemical analysis have made comparatively little progress among chemists (although the much weaker powder method is known and practiced by many of them). But the possibility of using the Laue method for crystallochemical analysis, especially by noncrystallographers, seems to have escaped attention.

Advantages of using the Laue method as a source of data for crystallochemical analysis

When single crystals are available, the Laue method offers an attractive way of returning to crystallochemical analysis. This method has the great practical advantage of requiring the simplest of apparatus provided an x-ray generator is already available. The required exposure time is extremely short, even in comparison with that needed for a powder photograph. This is because all orders of reflection from a plane (hkl) cooperate to produce a single spot on the photograph, a feature which the Laue method shares with no other x-ray diffraction method. Once the spots on a Laue photograph have been transferred to poles on a gnomonic projection, then from that point on, the interpretation makes use of the same crystallographic theory and, if necessary, the same crystallographic calculus which are needed in attempting to interpret the meaning of an array of faces on a crystal whose angular positions have been measured with the aid of an optical goniometer.

The morphological method, however, suffers from a tremendous disadvantage. Angular data are obtainable only from the relatively few planes that happen to occur on the surface of the crystal. This small sample is less than adequate for the determination of the correct gnomonic net that would provide proper indexing for all planes and would lead to a unique solution for the axial ratios of the crystal. It was partly this in-

[†] M. J. Buerger, Diffraction symbols. Chapter 3 of Physics of the solid state, S. Balakrishna, B. Ramachandra Rao, and M. Krishnamurthy, eds. (Academic Press, London, 1969) 27–42.

adequacy which gave rise to the tedious calculation in Fedorov's method and the artificial features of Barker's method.

In comparison with the inadequate data based on measurements of surface planes, the Laue method provides the corresponding angular data for *all* rational planes within the range of spacings permitted by the minimum wavelength of the x-ray source. Accordingly, when due account is taken of the range within which Laue spots contain only one order of reflection from the plane, the Laue method provides more than adequate data for fixing the crystallographically important geometrical characteristics of the crystal. These data lead directly to the correct values of the axial ratios of the true crystal cell. The ratios can be refined with the aid of reflections with large values of θ, especially if reflections from the back-reflection region are available, as they are, for example, in cylindrical Laue photographs, discussed in Chapter 9.

Data furnished by the Laue method

Friedel symmetry. Goniometric methods which locate surface planes supply only angular data, leading at best to the holohedral class of a crystal system. At worst, such methods fail even in this, for their information is metric and not symmetric in nature; for example, monoclinic arsenopyrite is metrically orthorhombic (indeed, the mineral is still assigned to the orthorhombic system by some mineralogy texts on this basis), monoclinic coesite is metrically hexagonal, and monoclinic kogarkoite is metrically rhombohedral. This uncertain symmetry information was one of the inadequacies of the early schemes of crystallochemical analysis.

By contrast, every Laue photograph provides the Friedel symmetry of the direction through the crystal traversed by the primary x-ray beam. By proper choice of directions for three Laue photographs (one along each direction symbolized by the three items of the general Hermann–Mauguin symbol) the complete Friedel symmetry of the entire crystal can be determined.

The Friedel symmetry is the maximum point-group symmetry information available by qualitative x-ray diffraction experiments. As was shown in Chapter 6, the 11 Friedel symmetries unequivocally determine the crystal system to which the crystal belongs, in addition to subdividing the tetragonal and isometrical systems into two categories of point groups each, and the hexagonal system into four categories of point groups. In this feature, the single-crystal x-ray diffraction methods provide exact information about crystal systems and crystal class that is unattainable by goniometric methods without an exhaustive study of crystal forms.

Crystal cell. The axial ratios can be transformed into cell dimensions, provided that the maximum voltage across the x-ray tube is known, by making use of the well-known reciprocal relation between voltage and minimum wavelength:

$$\lambda_{\min} V = \frac{hc}{e}$$

$$= 12.398 \quad \text{when } \lambda \text{ is expressed in Å and } V \text{ is expressed in kV.} \quad (1)$$

This provides the minimum wavelength for substitution in Bragg's equation. If the x-ray beam is along [001], for example, then Bragg's equation may be expressed in the form

$$\lambda_{\min} = 2d_{hk1} \sin \theta$$

$$= \frac{2 \sin \theta}{|{}^*\mathbf{t}_{hk1}|} \quad (2)$$

where

$$\mathbf{t}^*_{hk1} = h\mathbf{a}^* + k\mathbf{b}^* + \mathbf{c}^*. \quad (3)$$

Thus, the absolute magnitude of several reciprocal-lattice vectors $\mathbf{t}^*$ from the origin to the $hk1$ reciprocal-lattice points within the region of Laue spots containing only one order of reflection can be computed from

$$\mathbf{t}^*_{hk1} = \frac{2 \sin \theta_{hk1}}{\lambda_{\min}}. \quad (4)$$

The vector $\mathbf{t}^*_{hk1}$ can be decomposed into $\mathbf{t}'_{hk0} + \mathbf{t}'_{001}$. The first is the vector from the gnomonic-projection origin to gnomonic-net pole $hk1$ and the second is the distance of the plane of the gnomonic projection from the projection point (which is the radius of the sphere whose poles are projected on the gnomonic projection). Except for the correct scale, these vectors are available in the gnomonic projection and can be readily resolved graphically into their components on the right side of (3). The correct scale is provided by normalizing each set of components with the aid of (3) and (4), and selecting the maximum values of $|\mathbf{a}^*|$, $|\mathbf{b}^*|$, and $|\mathbf{c}^*|$ so determined.

These values of a^*, b^*, and c^* can be easily transformed into a, b, and c by standard formulas[†]. The values of the cell axes determined in this way are somewhat rough because (i) the minimum wavelength is known only

[†] M. J. Buerger. X-ray crystallography (Wiley, New York, 1942.) 360.

approximately, and (ii) the maximum values of $a*$, $b*$, and $c*$ of the available set approach, but usually do not attain, the true maximum possible.

Lattice type. Goniometric measurements give no direct information about the lattice type. An indirect method invokes the empirical law of Bravais, which states that the prominence of crystal forms is in the same order as their interplanar spacings. Since, for a given cell, spacings sequences differ for different lattice types, it is possible to compute, for each lattice type permitted by the symmetry, the sequence of interplanar spacings for that cell. The lattice type that best predicts the observed order of prominence of crystal forms is the one chosen as that of the crystal. This procedure has been successful in the hands of J. D. H. Donnay[†]. But the procedure normally requires judging the prominence of the several forms observed in a crystal by all the known examples, and so depends on more than mere goniometry. Furthermore, the tedious calculations[§] involved were part of the reason for the failure of the original Fedorov method of crystallochemical analysis.

But lattice type is evident on mere inspection of the gnomonic projection of a crystal. The basis for this is that all levels of a lattice are identical plane lattices. The space lattice is composed of a stack of plane lattices whose displacement from one level to another is the same for all nearest neighbors[‡]. The component of the displacement from a symmetry axis for a pair of neighboring planes fixes the lattice type. This displacement is zero if the lattice is primitive, and a submultiple of a cell translation if the lattice is not primitive. Since some Laue spots occur within a region where each spot has contributions from only the first level, the position within this region of a cell of the gnomonic net with respect to the origin fixes the lattice type by mere inspection.

Screw axes and glide planes. Screw axes are characterized by systematic absences in certain orders of reflections from planes (001), (010), (100), and (110). Such orders occur along central rows of the reciprocal lattice with very simple indices; reflections along such a row all contribute to a single Laue spot, so absences of separate orders cannot be observed.

Glide planes are characterized by systematic absences on zero levels

[†] J. D. H. Donnay and J. Mélon, Crystallography of lithium molybdotellurate. *Amer. Mineralog.* **21** (1936) 125–127.

[§] J. D. H. Donnay and J. Mélon, Haüy-Bravais lattice and other crystallographic data for sodium molybdo-tellurate. *Amer. Mineralog.* **18** (1933) 225–247.

[‡] M. J. Buerger. Elementary crystallography. (Wiley, New York, 1956.) 84–98.

normal to possible axes of the Friedel symmetries (which are the symmetry directions in the direct lattice). Laue photographs cannot record sets of reflections on zero levels. Thus, Laue photographs cannot easily give appropriate information about the presence of glide planes or screw axes in the space group. Some information may be gleaned from certain high-index reflections that have one zero index, but the number of such reflections available is hardly adequate to permit proper exploration for systematic absences in three nonequivalent sets of symmetry planes.

Diffraction symbol. From the foregoing discussion it follows that the Laue method can supply only part of the diffraction symbol, specifically the Friedel symmetry and the lattice type. Information about the rest of the space-group symmetry is generally unattainable by the Laue method. This is due entirely to the curious characteristic of the method that requires all orders of reflection to record in one and the same Laue spot.

Identification

Once the crystal system and axial ratios of a crystal are established, it is an easy matter to seek matching data in Donnay's *Crystal data*.[23–25] The determinative tables in this index are arranged first according to crystal system; within each of the six crystal systems the entries are listed according to axial ratio a/b for triclinic, monoclinic, and orthorhombic systems, according to c/a in the tetragonal and hexagonal systems, and according to a in the cubic system. The interaxial angles, the absolute lengths of the axes, ratios between other axes, and physical properties are used as secondary identification information. Some less comprehensive tables for the identification of minerals were supplied by Goldschmidt and Gordon[6] as early as 1928.

Donnay's tables are arranged in a form most useful for the Laue method because they were designed to be useful for data derived from optical goniometry as well as from x-ray diffraction. This objective accounts for the listing of the data by increasing axial ratio rather than by increasing axial translations. This manner of listing makes the tables ideal for crystallochemical analysis by the Laue method, in which the axial ratio is known more precisely than any translation.

It has been seen that with simple instrumentation and a single crystal, the Laue method provides a fairly precise value of the axial ratios, and somewhat rough values of the cell dimensions. It also supplies the crystal system, Friedel symmetry, and lattice type, thereby providing a very powerful method of identifying single crystals.

Significant literature

1. M. L. Frankenheim. System der Kristalle. Ein Versuch. (Grass, Barth und Comp., Breslau, 1842) 191 pages. [Besonderer Abdruck aus der 19ten Bandes der Nova Acta Acad. Nat. Cur. 191 pages.]
2. E. von Fedorow. Die Praxis in der krystallochemischen Analyse und die Abfassung der Tabellen für dieselbe. *Z. Kristallogr. Mineral.* **50** (1912) 513–575.
3. E. von Federow [unter Mitwirkung von D. Artemiere, Th. Barker, B. Orelkin, und W. Sokolov] Das Krystallreich. Mémoires de l'Académie des Sciences de Russie; Classe Physico-Mathématique XXXVI (1920) [der Akademievorgelegt am 26. Oktober 1911] lxiv + 1050 pages, and a volume of atlas with 149 sheets of diagrams.
4. T. V. Barker. Graphical and tabular methods in crystallography as a foundation of a new system of practice. (Thomas Murby, London, 1922) 152 pages.
5. A. K. Boldyrew. Principes de la nouvelle méthode de diagnose crystallographique de la matière. *Mém. Soc. Russe Min.* **53** (1925) 251–342. [Text in Russian, summary in French.]
6. V. Goldschmidt and S. G. Gordon. Crystallographic tables for the determination of minerals. Acad. Nat. Sci. Philadelphia, Special Publ. No. 2 (1928) 70 pages.
7. T. V. Barker. Systematic crystallography, an essay on crystal description, classification and identification. (Thomas Murby, London, 1930) 115 pages.
8. J. H. Haan. Kristallometrische Determineerings-methoden (Wolters, Groningen, 1932). Dissertation. [Text in Dutch, summary in German.]
9. J. D. H. Donnay and J. Mélon. Angles paramétriques de Barker dans une série cristalline homéomorphe. *Ann. (Bull.) Soc. Géol. Belg.* **57** (1933) B.39–B.52, Liège.
10. P. Terpstra, J. D. H. Donnay, J. Mélon, and W. J. van Weerden. Studies on Barker's determinative method of systematic crystallography. *Zeits. Krist.* (A)**87** (1934) 281–305.
11. J. D. H. Donnay and J. Mélon. Crystallo-chemical tables for the determination of tetragonal substances (laboratory compounds and artificial minerals) by means of their Barker angle and auxiliary physical properties. Johns Hopkins Studies in Geology, No. 11 (1934) 305–388.
12. A. K. Boldyrew and W. W. Doliwo-Dobrowolsky. Über die Bestimmungstabellen für Kristalle. *Zeits. Krist.* (A)**93** (1936) 321–367.
13. J. D. H. Donnay. Über die Bestimmungstabellen für Kristalle. *Zeits. Krist.* (A)**94** (1936) 410–412.
14. P. Terpstra and W. J. van Weerden. Studies on Barker's principle of simplest indices. *Z. Krist.* **A95** (1936) 368–382.
15. A. K. Boldyrew and W. W. Doliwo-Dobrowolsky. Bestimmungstabellen für Kristalle, Vol. I, Part 1: Einleitung. Tetragyrische Syngonie. (Leningrad and Moscow, 1937).
16. P. Terpstra and W. J. van Weerden. Over Barker's systematische Kristallographie. Nat. Wet. Tijdschr. Ghent, 20 (1938) 285.
17. Willem Jacob van Weerden. Algemeene Beschouwingen over Barker's "principle of simplest indices" (J. B. Wolters, Uitgevers-maatschappij, Groningen, 1938) 143 pages [with 6 page English summary.]
18. W. W. Doliwo-Dobrowolsky. Bestimmungstabellen für Kristalle. Vol. I, Part 2. Trigyrische und hexagyrische Syngonien allgemeine Ergänzung zu den mittleren Syngonien. (Leningrad and Moscow, 1939).
19. M. W. Porter and R. C. Spiller. Crystallo-chemical analysis. The Barker index at Oxford. *Nature* **144** (1939) 298–299.

20. W. G. Perdok. Kristallografische determineerings-methodes de Barker-methode. *Chemisch. Weekblad, Amsterdam,* **44** (1948) 202–207.

21. J. D. H. Donnay. Crystallochemical analysis. In "Physical methods of organic chemistry" (Wiley-Interscience, New York). Arnold Weissberger, editor.
    1st Ed., vol. 1 (1945) Chapter 13, 561–583.
    2nd Ed., vol. 1 (1949) Chapter 17, 1017–1039.
    3rd Ed., vol. 1, Part II (1960) Chapter 20, 1317–1346.

22. M. W. Porter and R. C. Spiller. The Barker index of crystals. A method for the identification of crystalline substances. (W. Heffner & Sons Ltd., Cambridge, England).
    Vol. I (1951). Crystals of the tetragonal, trigonal and orthorhombic systems.
    Part 1. Introduction and tables. 120 pages and 9 long tables.
    Part 2. Crystal descriptions [tables].
    Vol. II (1956). Crystals of the monoclinic system.
    Part 1. Introduction and tables. 56 pages and 8 long tables.
    Part 2. Crystal descriptions M.1 to M.1800 [tables].
    Part 3. Crystal descriptions M.1801 to M.3572 [tables].
    Vol. III (1969). Crystals of the anorthic system.
    Part 1. Introduction and tables. 50 pages and 11 tables.
    Part 2. Crystal descriptions [tables and atlas of configurations].

23. J. D. H. Donnay, Werner Nowacki and Gabrielle Donnay. Crystal data. Geological Society of America Memoir 60 (1954) 719 pages.

24. J. D. H. Donnay, Gabrielle Donnay, E. G. Cox, Olga Kennard, and Murray Vernon King. Crystal data determinative tables, second edition. Am. Crystallogr. Assoc. Monograph Number 5 (1963) 1302 pages.

25. J. D. H. Donnay, Helen M. Ondik, Sten Samson, Mary E. Mrose, Olga Kennard, Murray Vernon King, and William R. Bozman. Crystal data determinative tables, third edition.
    Vol. I (1972) Organic compounds IX + 839 pages.
    Vol. II (1973) Inorganic compounds IX + 1696 pages.

Chapter 8

Planes and zones

In the foregoing chapters the plotting of projections, either stereographic, gnomonic, or stereognomonic, has been explained and the basic constructions that deal with simple crystallographic problems such as the drawing of zones and the measurement of angles have been treated. The question remains, however, how to make use of projections. This question was answered long ago by crystallographers. One of the basic things that can be determined from a projection is the actual symmetry of the crystal. Another question is the determination of the orientation of the crystal with respect to the x-ray beam. Both questions are also intimately related to another basic crystallographic problem, the selection of the correct set of crystal axes and the indexing of the poles. To solve such problems the researcher must rely upon the crystallographic calculus that was highly developed by classical crystallographers who studied crystal morphology by optical goniometry. In what follows, this calculus will be used extensively.

Basic crystallographic relations

As a consequence of the essential three-dimensional regularity of crystals, two important morphological laws apply to crystals: the law of rational indices and the law of zones. The first states that the indices of a crystal face are integers, usually small integers. The second law states that all the faces of a crystal tend to fall into a strictly limited number of zones.

Zone is an old morphological name for a rational direction; faces in a zone are parallel to a common rational direction called the *zone axis*.

These two morphological laws are of the greatest importance in the interpretation of the Laue photographs because such a photograph is only a special projection of the polar sphere discussed in Chapter 9. The first law is important for the indexing of the spots on the Laue photograph; the second law, for the deduction of the symmetry of the crystal. We now examine some important relations between crystal planes and zones.

Rational planes and rows. A lattice is defined by three translations[†] **a**, **b**, **c**; these correspond to three directions in the space, called the crystal axes. The coordinates of a lattice point in a primitive lattice are ua, vb, wc. The three coefficients u, v, w are integers and are known as the indices of the lattice point.

Any lattice point may be chosen as an origin. The origin O of the lattice and any lattice point P determine a lattice row. This lattice direction OP is defined by

$$\overrightarrow{OP} = u\mathbf{a} + v\mathbf{b} + w\mathbf{c}. \tag{1}$$

If u, v, and w are mutually prime, the lattice point P is the nearest to the origin in the direction OP; and these three integers, placed in brackets thus $[uvw]$, are used to represent the indices of the corresponding lattice row. The three rows $[100]$, $[010]$, and $[001]$ are the crystal axes.

Any three noncollinear lattice points define a rational plane. Of all the rational planes that occur in a lattice, only those that pass through the origin of the lattice are of interest in the Laue method. To determine the indices of such rational planes, let us consider two lattice points uvw and $u'v'w'$ that, with the origin, constitute a noncollinear set. The indices of the corresponding lattice rows are $[uvw]$ and $[u'v'w']$. These two rows determine a rational plane whose indices are (hkl), passing through the origin of the lattice. The equation of this plane can be written as

$$h\frac{x}{a} + k\frac{y}{b} + l\frac{z}{c} = 0. \tag{2}$$

Here, x, y, and z are absolute coordinates, while a, b, and c are the lengths of the cell edges. A point of this plane that also belongs to the lattice row

---

[†] If the lattice is primitive, it is most convenient to choose the three shortest translations. These define the reduced primitive cell.

[uvw] has, as coordinates,

$$x = ua, \tag{3}$$

$$y = vb, \tag{4}$$

$$z = wc. \tag{5}$$

Substituting these values in (2), there is obtained

$$hu + kv + lw = 0. \tag{6}$$

This equation states the condition[†] for a lattice row to be contained in a lattice plane.

In a similar way, for the second lattice row [$u'v'w'$],

$$hu' + kv' + lw' = 0, \tag{7}$$

must hold. The two equations (6) and (7) contain three numbers h, k, and l. The relation between these must be

$$\frac{h}{vw' - wv'} = \frac{k}{wu' - uw'} = \frac{l}{uv' - vu'}. \tag{8}$$

Substituting in (2) the values of h, k, and l for their proportional equivalents given in (8), there is finally obtained

$$(vw' - wv')\frac{x}{a} + (wu' - uw')\frac{y}{b} + (uv' - vu')\frac{z}{c} = 0. \tag{9}$$

Certain crystal planes have special significance. An *axial plane* is the plane defined by any two of the crystal axes. There are three axial planes in a crystal, namely (100), (010), and (001). A *parametral plane* is the plane all of whose indices are unity, specifically (111).

Of the general crystal planes defined by (9) not all have the same importance. Those that have simple indices are the most important from a crystallographic point of view. Simple indices, as meant here, are only 0, 1, or $\bar{1}$. These crystal planes are normally planes of high density of lattice points.

Tautozonal planes. The intersection of two lattice planes (hkl) and ($h'k'l'$) determines a lattice row [uvw]. Since [uvw] is contained in the

---

[†] The expression $hu + kv + lw$ is also the scalar (or "inner") product of two vectors, one of these, $u\mathbf{a} + v\mathbf{b} + w\mathbf{c}$, in direct space; the other, $h\mathbf{a}^* + k\mathbf{b}^* + l\mathbf{c}^*$, in reciprocal space (since $\mathbf{aa}^* = 1$ and $\mathbf{ab}^* = 0$, etc.)

crystal planes, (6) and (7) show that the following must hold:

$$hu + kv + lw = 0;$$

$$h'u + k'v + l'w = 0.$$

These require that

$$\frac{u}{kl' - lk'} = \frac{v}{lh' - hl'} = \frac{w}{hk' - kh'}.$$

Three crystal planes (hkl), $(h'k'l')$, and $(h''k''l'')$ are in the same zone when their intersection is a common crystal row $[uvw]$, for which it is necessary that the three equations

$$h\frac{x}{a} + k\frac{y}{b} + l\frac{z}{c} = 0,$$

$$h'\frac{x}{a} + k'\frac{y}{b} + l'\frac{z}{c} = 0, \tag{10}$$

$$h''\frac{x}{a} + k''\frac{y}{b} + l''\frac{z}{c} = 0$$

have a common solution. This is attained when the determinant of the coefficients is zero; that is,

$$\begin{vmatrix} h & k & l \\ h' & k' & l' \\ h'' & k'' & l'' \end{vmatrix} = 0. \tag{11}$$

The planes that belong to the same zone are said to be *tautozonal*.

The crystal planes with simple indices define nine important zones in the crystal, namely,

$$\begin{array}{cccc} [100] & [110] & [101] & [011] \\ [010] & [1\bar{1}0] & [\bar{1}01] & [0\bar{1}1] \\ [001] \end{array}$$

Tautozonality and zone development. In (11) we have seen that the determinant of the coefficients of the indices of three planes that belong to one zone is zero. The common crystal direction $[uvw]$ is known as the zone axis. The basic condition for a crystal plane (hkl) to belong to the zone $[uvw]$ is that the scalar product of the indices of the plane and zone must be zero, as given in (6).

If two crystal planes $(h_1k_1l_1)$ and $(h_2k_2l_2)$ are in a zone, any other plane

Table 1

*Development of planes according to Goldschmidt's law
of complication, as described by Federov symbols*

Degree of development of the zone	Fedorov symbols for planes present in the zone								
0	10								01
1	10			11					01
2	10		21	11		12			01
3	10	31	21	32	11	23	12	13	01
.	.	.	.	.	.	.	.	.	.

lying in the same zone and located between the two planes has for its symbol

$$(nh_1 + mh_2, \qquad nk_1 + mk_2, \qquad nl_1 + ml_2)$$

where n and m are positive or negative integers. These two numbers n and m are called Fedorov zonal symbols.[4,5] This relation has foremost importance in understanding the succession of planes in a zone; the succession is known as the *development of the zone*. Let us assume (100) and (010) to be planes that limit a region of the zone [001]. By giving n and m the value 1, the plane (110) is obtained. If $n = 2$ and $m = 1$, the plane (210) is defined; this is situated between (100) and the previously obtained plane (110). If $n = 1$ and $m = 2$, the plane (120) is obtained, located between (110) and (010). By this iterative process, further planes are obtained.

The preceding sequence can be also obtained in the following way. The successive planes in the zone have indices that are sums of those previously generated. This notion was described by Goldschmidt[3,5] in what he called the *law of complication*[†] of a zone. It is summarized in a schematic way in Table 1. Goldschmidt sought the complication of planes in a zone as the consequence of successive generations. The starting planes constitute the zero generation, the first face (110) is the first generation, and so on. It can be seen that the generation coincides with the highest value of m or n. A zone that follows the development predicted by Goldschmidt's law is known as a *normal zone*. Any other zone that does not follow the law is called an *abnormal zone*.

Let us examine more closely some details of the zone development. If, between the two end planes, there are no other crystal planes shown in the zone, the zone is said to be of zero-degree development. The plane that is

[†] It might better be called "the zonal interpolation law".

found by summing the indices of the end planes is called the dominant plane (or face). The zone that contains only the two end faces and the dominant is called the zone of first-degree development, and so on. In a zone of first-degree development the Fedorov zonal symbol is 11. For a zone showing second-degree development the new Fedorov zonal symbols are 21 and 12. Analogously, for a zone that shows third-degree development, the new zonal symbols are 31, 32, 23, 13, and so on. If we extrapolate this concept to the zone of order zero, the zonal symbols are 10 and 01.

The zonal symbols also follow the law of complication. The crystal planes that appear in the zones, given by zonal symbols, depend on the degree of the zone as shown in Table 1. The law of complication is general and applies to the whole zone or parts of a zone.

It can be observed that, according to the complication theory, zones should be symmetrical in development. The symmetry is not crystallographic symmetry, however. What apparent symmetry is observable occurs only because the existence of a lattice provides a symmetry between the integers h and k, for example.[18]

Very few crystals show normal zones in their morphological development. Also, in the Laue photographs, where each crystal plane is potentially able to give a Laue spot, normal zones are not always observed because the intensity of a given reflection can be zero for various reasons. Even with this limitation, however, the law has a practical application in the study of zones in a Laue photograph, as is discussed in Chapter 9.

The normal pattern of the projection

The various projections for interpreting Laue photographs can be interpreted once the system of the nine zones has been identified. This is equivalent to having recognized the symmetry of the crystal and having identified the proper set of crystal axes.

The basic features of the crystal can be readily deduced from the projection obtained when the projection plane is perpendicular to the c axis. In this projection, called the *normal pattern*, axial planes (100) and (010) are parallel to the c axis, so their normals are perpendicular to that axis and therefore lie on the plane of the primitive circle. It is customary to assume the pole (010) as the origin of the angular coordinate ϕ; this direction is oriented toward the east in the projection. The (100) pole is now situated in the south region of the projection. The (001) pole is in the neighborhood of the center of the projection (Fig. 1). It can be easily seen that [001] lies in the center of the projection, and that [$\bar{1}$00] lies toward the top and [0$\bar{1}$0] toward the upper left part of the projection.

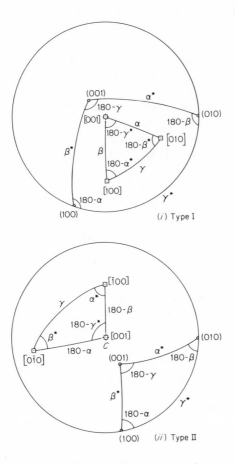

Fig. 1. Normal patterns of the stereographic projection of triclinic crystals.

The poles of the axial planes define three important angles, namely α^*, β^* and γ^*, the axial angles of the reciprocal lattice. The direct axial angles α, β, γ can be read from the intercepts of the zone axes [100], [010], and [001]. In Fig. 1, the angles $180° - \alpha$, $180° - \beta$, and $180° - \gamma$ are shown. Recall that the intercept of the zone axis is the pole at 90° from the circle defining the zone. Some relations between the direct and reciprocal axial angles and the spherical triangles formed by the poles of the axial planes and the axial rows are also shown in the same figure. The normal pattern of the projection of a crystal has the important advantage of allowing the determination of the crystal symmetry by inspection. The normal pattern for each crystallographic system has certain features to be discussed that make the identification of the projection unique. These features are the following.

For the triclinic crystal, in general, no zone axis coincides with a pole of a crystal plane. All the zones intersect at angles different than 90°.

For the monoclinic crystal, in general, only one main zone has its axis in coincidence with a pole of a crystal plane. In the first setting, this pole is (001) and the zone is [001]; in the second setting, which is the standard one used for monoclinic crystals in classical crystallography, the pole is (010) and the zone axis is [010]. There is only one set of zones perpendicular to another zone. In the second setting, these are zones that contain (010); the singular zone is [010]. All other zones are inclined.

For the orthorhombic crystal, three main zones are mutually perpendicular. These are the [100], [010], and [001] zones. Each of these zone axes coincides in the projection with a given pole of a crystal plane, namely, (100), (010), and (001). These three sets of zones are mutually perpendicular, but there are no specialized angles between the general poles in the zones.

For the tetragonal crystal, the same conditions prevail as for the orthorhombic crystal. However, for the tetragonal crystal, one zone, [001], has the poles (100), (110), (010), ($\bar{1}$10), ($\bar{1}$00), ($\bar{1}\bar{1}$0), (0$\bar{1}$0), (1$\bar{1}$0) at 45° intervals.

For the cubic crystal, the basic feature of orthogonality of zones is present and, in addition, all the three main zones have poles at 45° to each other.

For the hexagonal crystal, one of the zones, [0001], has poles at 30° to each other. Three main zones can be recognized perpendicular to [0001] at 60° to each other.

Interpretation of the symmetry from certain features of a Laue photograph

If the Laue photograph is taken with the *c* axis of the crystal parallel to the x-ray beam, the interpretation of the projection is straightforward when the ideas given in the preceding section are applied. However, this special direction of incidence is highly improbable when a Laue photograph of a crystal of unknown orientation is obtained. The projection is usually in a general direction, so a method is needed to identify the symmetry of the crystal from the projection of the Laue photograph. The method is concerned with identifying the system of the nine zones of the crystal.

The first problem is to recognize in the projection the axial planes and the axial zones. As a clue, we may use the appearance of blank areas around some poles, a feature that was discussed in Chapter 5. These poles have simple indices. Also, well-populated zones show that the corresponding zone axes are crystallographic directions with simple indices. With this in mind, it is not very difficult to recognize the system of the nine zones, and only a few trials are ordinarily needed to find the correct set.

In the determination of the symmetry of the crystal from the projection of the Laue photograph, however, some complication arises, due to the fact that the Laue photograph gives all the possible crystal angles without regard to the development of the faces. As a consequence, the projection corresponds to the centrosymmetric Friedel classes of the corresponding crystal system, as discussed in Chapter 6. In analyzing the projection of the crystal faces, the handicap is avoided because the relative development of faces (hkl) and $(\bar{h}\bar{k}\bar{l})$ may be observable; if so, this knowledge is used in determining the real symmetry class of the crystal under study.

In the interpretation of a Laue diagram, we must rely on the intensities of the Laue reflections in the deduction of the Friedel symmetry. The intensity of a spot is proportional to the sum of the squares of the structure factors of the reflections recorded in the spot, so that the intensities of spots corresponding to symmetrically equivalent crystal planes are equal. Hence, the difference between Friedel classes of the same crystal system must be sought in the difference in the intensity of reflection from crystal planes that would be equivalent in the holohedral crystal form. These circumstances are included in Table 2. If we disregard the reflections due to characteristic radiation (which can easily be identified by their intensity and special position), the intensities of symmetrically equivalent Laue spots should be equal. This is not absolutely true for unsymmetrical photographs since the intensity of the continuous spectrum is not constant and the polarization factor plays an important role in the observed intensity. However, in any particular small region of the photograph the principle holds very approximately, and therefore the criterion of the intensity can be used.

In the Laue method, we are always dealing with what is called "overdevelopment" in crystal goniometry; that is, we have reflections corresponding to too many faces, and although the selection of the appropriate ones is therefore sometimes cumbersome, it is by no means difficult. We shall give here some rules for the proper use of the information obtained from a Laue photograph in order to determine the Friedel symmetry of the crystal.

Orthorhombic crystals. The identification of the principal zones does

Table 2

Determination of Friedel class by reflection-intensity relations

Crystal system	Minimum equivalence required by crystal system	Friedel class	Distinctive relations for Friedel class
Triclinic		$\bar{1}$	$hkl = \bar{h}\bar{k}\bar{l}$ only
Monoclinic	1st setting: $hkl = \begin{cases} \bar{h}k\bar{l} \\ \text{or } hk\bar{l} \end{cases}$	$\dfrac{2}{m}$	$hkl = \bar{h}k\bar{l} = hk\bar{l} = \bar{h}k\bar{l}$
	2nd setting: $hkl = \begin{cases} \bar{h}k\bar{l} \\ \text{or } h\bar{k}l \end{cases}$		$hkl = \bar{h}k\bar{l} = h\bar{k}l = \bar{h}\bar{k}\bar{l}$
Orthorhombic	$hkl = \bar{h}\bar{k}l = \begin{cases} hk\bar{l} = \bar{h}\bar{k}\bar{l} \\ \text{or } h\bar{k}l = \bar{h}k\bar{l} \end{cases}$	$\dfrac{2}{m}\dfrac{2}{m}\dfrac{2}{m}$	$hkl = \bar{h}\bar{k}l = hk\bar{l} = \bar{h}\bar{k}\bar{l}$ $= \bar{h}\bar{k}\bar{l} = h\bar{k}\bar{l} = \bar{h}k\bar{l} = h\bar{k}l$
Tetragonal	$hkl = k\bar{h}l = \bar{h}\bar{k}l = k\bar{h}l$	$\dfrac{4}{m}$	$hkl \neq \bar{h}k\bar{l}$
		$\dfrac{4}{m}\dfrac{2}{m}\dfrac{2}{m}$	$hkl = h\bar{k}l$

Crystal system		Point group		
Hexagonal	$hkil = ihkl = kihl$	$\bar{3}$		$hkil \neq khil$
		$\bar{3}\,\dfrac{2}{m}$	$hkil = khil$	
		$\dfrac{6}{m}$	$hkil = \bar{k}\bar{i}\bar{h}l = hki\bar{l}$	$hkil \neq khil$
		$\dfrac{6}{m}\,\dfrac{2}{m}\,\dfrac{2}{m}$	$hkil = \bar{k}\bar{i}\bar{h}l = hki\bar{l}$ $hkil = khil$	
Cubic	$hk\bar{l} = \bar{l}hk = k\bar{l}h$ $h\bar{k}\bar{l} = l\bar{h}\bar{k} = k\bar{l}\bar{h}$ $\bar{h}k\bar{l} = \bar{l}\bar{h}k = \bar{k}lh$ $\bar{h}\bar{k}l = \bar{l}\bar{h}k = \bar{k}\bar{l}h$	$\dfrac{2}{m}\,\bar{3}$	$hkil = \bar{h}\bar{k}l = \bar{l}hk$	
		$\dfrac{4}{m}\,\bar{3}\,\dfrac{2}{m}$	$hkl = hkl = h\bar{l}k = l\bar{h}k$ $hkl = hlk = khl = lkh$	$hkl \neq hlk = khl = lkh$

not present any particular problem and therefore the determination of the symmetry of any orthorhombic crystal is in some ways the simplest case. It is necessary to identify the three mutually perpendicular zones, and to check that a crystal pole coincides with the intersection of the corresponding zone axis. When this has been done, the axial zones have been identified. In order to exclude the tetragonal or hexagonal symmetry, we must be sure that no poles are located in any of the three zones at 45° or 60° (or 30°) to each other.

However, a problem arises when it is necessary to distinguish properly between the three crystallographic axes. In classical mineralogy these axes are usually selected so that $c < a < b$. An appropriate angular selection has to be made by identifying, in the corresponding [100], [010], and [001] zones, the angular relations

$$45° < (110) \wedge (010) < (101) \wedge (100) < (011) \wedge (010).$$

These relations give the appropriate selection of the three crystallographic axes. The identification of such angles between important poles in the projection solves, in one stroke, the right setting and, accordingly, the right selection of the crystal axes.

In applying this method, a most useful kind of photograph has been found to be the cylindrical Laue photograph. This method of recording is treated in Chapter 9. In Fig. 2 a Laue photograph of NH_4NO_3 IV is shown. Its stereographic projection is given in Fig. 3. The three axial planes and the three axial zones are identified, showing the existence of three mutually perpendicular zones. The symmetry is orthorhombic. This case and the following examples show that the symmetry of the crystal can be determined with the aid of only one cylindrical Laue photograph.

Monoclinic crystals. In this case, a new problem arises. First, the main zone 90° from an important pole must be identified. For the traditional second setting of monoclinic crystals, this corresponds to the [010] zone and the (010) plane. After this has been done, the proper selection of the main faces in such a zone must be achieved in order to have the right selection of the crystal axes. Since we are dealing here with overdevelopment, some rules are necessary.

Because of the limitation imposed on the choice of axial directions and lengths, overdevelopment does not give much trouble in the more symmetrical systems, as we shall see. In the monoclinic system, however, it can lead to situations in which it is quite difficult to be sure that the appropriate setting has been found. This ambiguity is even more serious in the triclinic system.

Before going further, let us remember that the appropriate setting is reached when [010] and (010) are in coincidence in projection, and that $a > c$. Also, the angle $\beta = (100) \wedge (001)$ must be obtuse, so $\beta^* = 180° - \beta$ in the projection must be acute. With this convention, the positive ends of a and c meet at an obtuse angle.

The selection of (100) and (001) is made between the important poles in the zone by taking into account the following rules.

1. When several choices exist, the choice that gives $(100) \wedge (001)$ nearest to 90° must be selected.

2. Between poles (100) and (001) the first important reflection (or pole) can be selected by taking into account that $(100) \wedge (101) > (101) \wedge (001)$.

3. The angle $(100) \wedge (101)$ should be approximately 45°. Of course, for crystals having very different a and c lengths, a criterion for the selection of the important faces is to select only those marked by the confluence of many important zones.

As an example, a Laue photograph of triglycine sulfate is given in Fig. 4 and the corresponding projection in Fig. 5. The projection is again fully interpreted, showing the monoclinic symmetry of the crystal.

Triclinic crystals. In the case of the triclinic crystal no specialization in the axial set occurs, and few general rules can be given. A rule used by classical mineralogists is $c < a < b$. If the three interaxial angles are not to be treated differently, it must be recognized that α, β, γ in the crystal may all be either obtuse, corresponding to a type-II cell, or acute, corresponding to a type-I cell.

A crystal of malonic acid has been taken as example. Figures 6 and 7 show a Laue photograph and the corresponding stereographic projection. The projection shows one selection of the axial planes for a triclinic crystal.

Tetragonal crystals. The tetragonal system is characterized by the presence of a set of four important zones [100], [110], [010], and [1$\bar{1}$0] at 45° intervals, and each perpendicular to another zone. This fifth zone is [001], so that this zone is easily determinable. The selection between the zones [100] (and [010]), and [110] (and [1$\bar{1}$0]) must be made by identifying the principal faces in the zones [100] and [110], respectively, namely, by identifying the poles (101) and (111) in the respective zones.

The right selection can be made by utilizing the requirement that any tetragonal crystal satisfies the following: $(001) \wedge (101) < (001) \wedge (111)$. The poles (101) and (111) will be marked in the projection by the convergence of numerous zones, and therefore the appropriate selection can be made unambiguously.

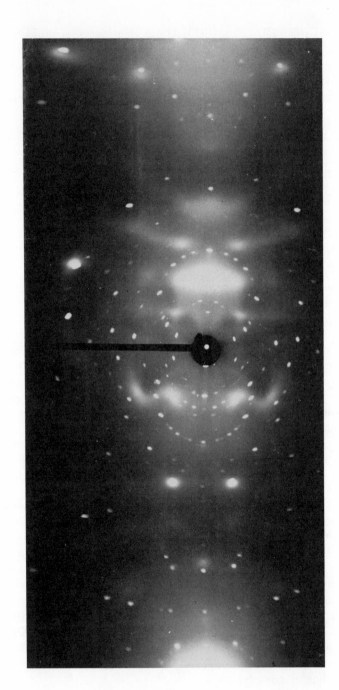

Fig. 2. Cylindrical Laue photograph of an orthorhombic crystal, NH₄NO₃ IV.

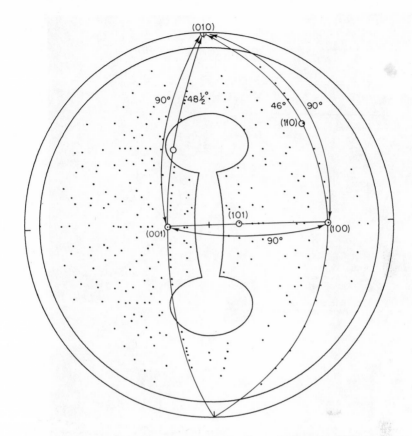

Fig. 3. Stereographic projection of the Laue photograph of Fig. 2.

Figure 8 shows a Laue photograph of a tetragonal crystal of pentaerythritol and Fig. 9 its stereographic projection.

Ambiguity in distinguishing between Friedel symmetries $4/m$ and $4/m\,2/m\,2/m$ can be avoided by applying the intensity criteria from Table 2, specifically that either

$$I(hkl) \neq I(khl) \qquad \text{for} \quad \frac{4}{m}$$

or

$$I(hkl) = I(khl) \qquad \text{for} \quad \frac{4}{m}\frac{2}{m}\frac{2}{m}$$

Fig. 4. Cylindrical Laue photograph of a monoclinic crystal, triglycine sulfate.

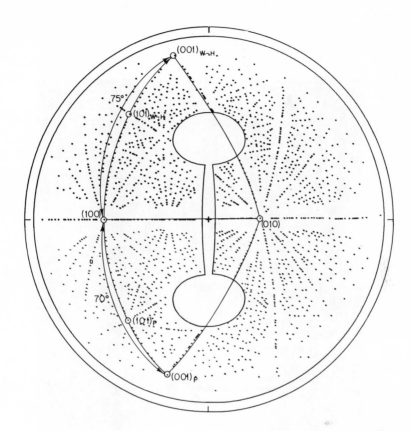

Fig. 5. Stereographic projection of the Laue photograph of Fig. 4 (W–H, Wood–Holden; P, Pepinsky).

Hexagonal and trigonal crystals. The syngonal characteristic of hexagonal and trigonal systems is the presence of an important pole at 90° from a zone with intrazonal pole angles of 30° or 60° (according as they correspond with hexagonal or trigonal crystals, respectively).

Hexagonal crystals may present problems analogous to those belonging to tetragonal symmetry; the main differences are that here the equivalent main zones are 60° or 120° apart. The selection between the zones $[01\bar{1}0]$ and $[11\bar{2}0]$ can be made in a way similar to that used in the tetragonal case, specifically:

$$(10\bar{1}1) \wedge (0001) < (11\bar{2}1) \wedge (0001)$$

The distinction between the $6/m$ and $6/m\ 2/m\ 2/m$ groups can be

Fig. 6.   Cylindrical Laue photograph of a triclinic crystal, malonic acid.

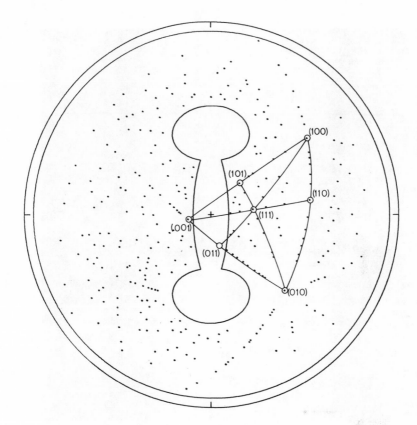

Fig. 7. Stereographic projection of the Laue photograph of Fig. 6 (system of axes is left-handed).

made by taking into account a possible difference in intensity of reflections from the reflections $hkil$ and $ikhl$, as seen in Table 2.

Further trouble can appear when considering a trigonal crystal because, due to overdevelopment, it is goniometrically identical with a hexagonal one. This fact would render impossible the discrimination between them by the Laue method if the intensity distribution of the reflections along non-equivalent zones were not taken into account. In a trigonal crystal, only three zones containing (0001) are equivalent by symmetry (and therefore, by intensity distribution) compared with the six main zones of the hexagonal crystal.

Figures 10 and 11 correspond to a Laue photograph and its stereographic projection, respectively, of a trigonal crystal, quartz (Friedel class $\bar{3}\,2/m$).

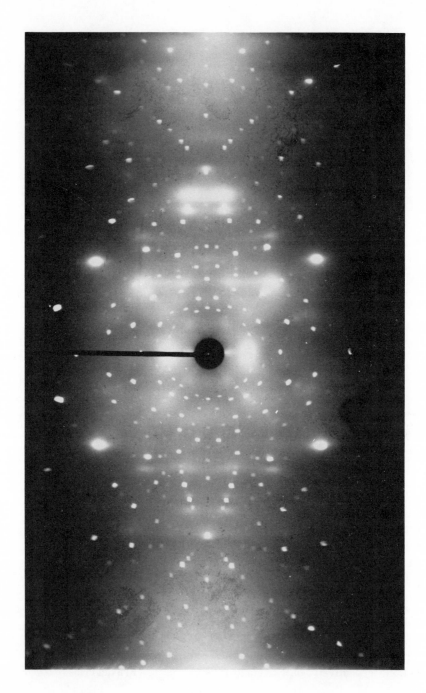

Fig. 8.   Cylindrical Laue photograph of a tetragonal crystral, pentaerythritol.

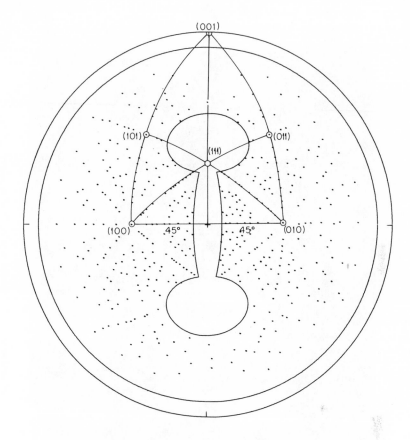

Fig. 9. Stereographic projection of the Laue photograph of Fig. 8.

Figure 11 shows the appearance of zones that have equal intensity distribution in the Laue photograph, spaced at 120°; halfway between these are interleaved a similar but nonidentical set of such zones, the characteristic feature of the fundamental domain of this system.

Cubic crystals. The cubic system is characterized by the existence of three main zones at right angles to one another; in each zone important nonequivalent poles occur alternately at 45° intervals. The zones correspond to [100], [010], and [001]. However, the zones [110], [011], and [101] are also orthogonal and may appear to determine a reference system; this must be distinguished from the true reference system. The difference between zones [100] and [110] is simple; the main angles in the zone [100]

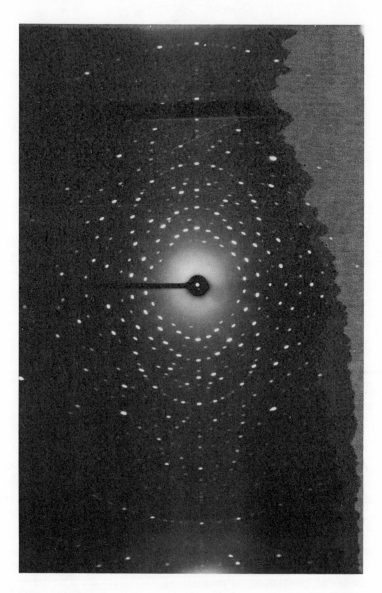

Fig. 10.   Cylindrical Laue photograph of a trigonal crystal, quartz.

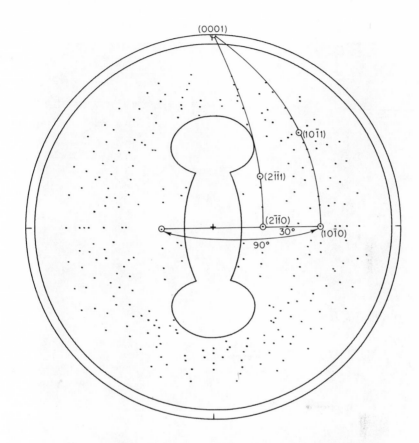

Fig. 11. Stereographic projection of the Laue photograph of Fig. 10.

are 45°, corresponding to (100) $\wedge$ (110). In the zone [110], however, the angles are of 54° 44′ 08″ and 35° 15′ 52″, corresponding respectively to [100] $\wedge$ [111] and [111] $\wedge$ [110]. The identification of the two most important zones is therefore easy: it is necessary to identify the sequence of interpolar angles in each zone. NaCl has been taken as an example in Figs. 12 and 13. The presence of the 45° rhythm identifies zones [100]; the alternation of poles at 55° and about[†] $35\frac{1}{2}$° shows that the measured zone is [110]. Since there are only three [100] zones, the determination of the crystal axes is easy. Hexamine, illustrated in Figs. 14 and 15, gives an example of a complicated case because, due to the orientation of the crystal,

[†] The error in the projection is of the order of $\pm\frac{1}{2}$°.

Fig. 12. Cylindrical Laue photograph of a cubic crystal, NaCl.

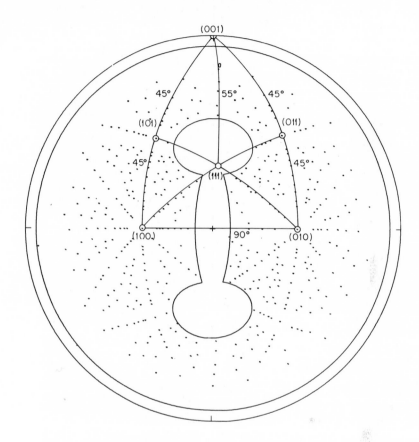

Fig. 13. Stereographic projection of the Laue photograph of Fig. 12.

many of the important reflections were located in the blind areas and therefore were not recorded. However, the identification of the principal zones can be made and, accordingly, the crystal symmetry and orientation of the crystal with respect to the x-ray beam can be determined.

The two possible cubic Friedel classes are $2/m\,\bar{3}$ and $4/m\,\bar{3}\,2/m$. Due to the overdevelopment given by the Laue photograph, these classes are goniometrically indistinguishable. The distinction between them can be made through the study of the intensities of the reflections in certain important zones, as shown in Table 2.

Axial ratio and the crystal lattice

The determination of the symmetry of the crystal through the system of the nine zones implies that the crystal planes with simple symbols have

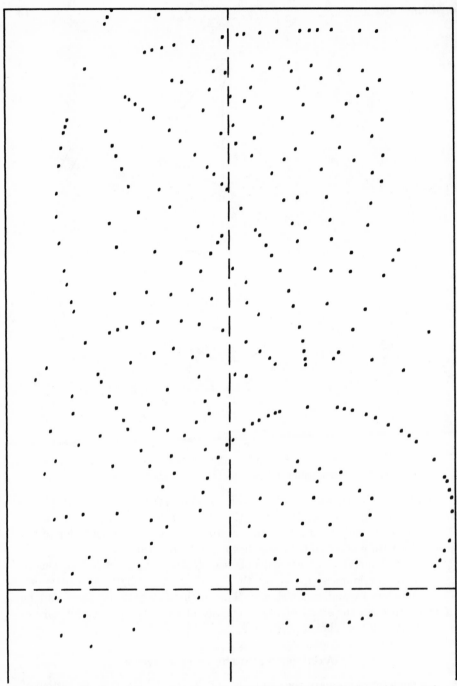

Fig. 14. Replica of the spots of a cylindrical Laue photograph of a cubic crystal, hexamine.

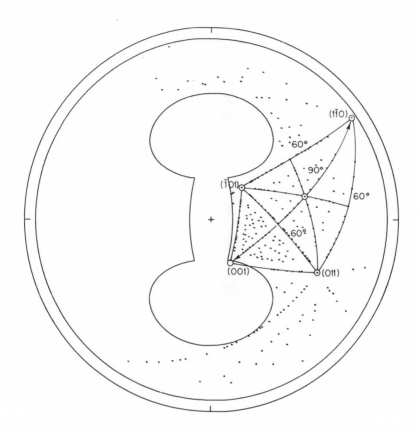

Fig. 15. Stereographic projection of the spot locations of Fig. 14.

been identified. Once this has been done, it is possible to determine $a:b:c$, known as the axial ratio. However, from the measurement of angles alone the absolute magnitudes of a, b, and c cannot be determined. Since there are an infinite number of $a':b':c'$ values that satisfy the axial ratio, it is customary to divide the axial ratio by b, as follows.

$$a':b':c' = \frac{a}{b} : 1 : \frac{c}{b}. \tag{12}$$

The new ratio is commonly expressed in crystal morphology as $a':1:c'$ where a' and c' now represent the ratios a/b and c/b, respectively, rather than the actual values of the respective lattice translations.

The system of the nine zones shown in Fig. 16 defines six important angles useful for the calculations of the axial ratio in the following way.

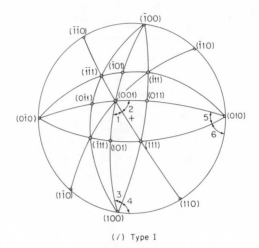

(*i*) Type I

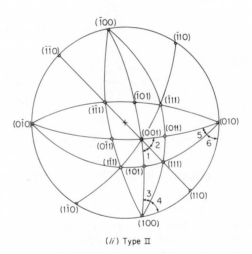

(*ii*) Type II

Fig. 16. The system of the nine zones and auxiliary angles for triclinic crystals of type I and type II.

$$\frac{a}{b} = \frac{\sin \angle 1}{\sin \angle 2}; \quad \frac{c}{b} = \frac{\sin \angle 3}{\sin \angle 4}; \quad \frac{c}{a} = \frac{\sin \angle 5}{\sin \angle 6}. \tag{13}$$

In general, the auxiliary angles $\angle 1, \ldots, \angle 6$ are not known, but they can be easily calculated from the angles between the poles of the important planes. The following relations exist between the auxiliary angles and the axial angles of the direct lattice.

$$\sphericalangle 3 + \sphericalangle 4 = 180° - \alpha,$$
$$\sphericalangle 5 + \sphericalangle 6 = 180° - \beta, \tag{14}$$
$$\sphericalangle 1 + \sphericalangle 2 = 180° - \gamma.$$

The axial ratio can also be calculated when the pole of the fundamental plane (111) and the intercepts of the axial directions have been identified in the projection. Let us define

$$(111) \wedge [100] = N,$$
$$(111) \wedge [010] = M, \tag{15}$$
$$(111) \wedge [001] = S.$$

Then

$$\frac{a}{b} : 1 : \frac{c}{b} = \frac{\sec N}{\sec M} : 1 : \frac{\sec S}{\sec M}. \tag{16}$$

The calculation of the axial ratio can be made in a simple way by using the auxiliary angles. The method in all its generality is required for triclinic crystals, but the introduction of higher symmetries implies reductions in the equations used. The specific forms of the method for obtaining the axial ratio for each one of the crystallographic systems are as follows.

Triclinic: The parametric ratio is $a:1:c$. The lack of symmetry results in $\alpha \neq \beta \neq \gamma \neq 90°$. The interaxial angles can be computed from

$$\alpha = 180° - (\sphericalangle 3 + \sphericalangle 4),$$
$$\beta = 180° - (\sphericalangle 5 + \sphericalangle 6), \tag{17}$$
$$\gamma = 180° - (\sphericalangle 1 + \sphericalangle 2).$$

The angles $\sphericalangle 1, \sphericalangle 2, \ldots, \sphericalangle 6$ cannot be measured directly and have to be calculated from angles between crystal poles in the following way.

$$\frac{a}{b} = \frac{\sin \sphericalangle 1}{\sin \sphericalangle 2} = \frac{\sin (100) \wedge (110)}{\sin (010) \wedge (110)} \frac{\sin (010) \wedge (001)}{\sin (100) \wedge (001)},$$

$$\frac{c}{b} = \frac{\sin \sphericalangle 3}{\sin \sphericalangle 4} = \frac{\sin (001) \wedge (011)}{\sin (010) \wedge (011)} \frac{\sin (100) \wedge (010)}{\sin (100) \wedge (001)}, \tag{18}$$

$$\frac{c}{a} = \frac{\sin \sphericalangle 5}{\sin \sphericalangle 6} = \frac{\sin (001) \wedge (101)}{\sin (100) \wedge (101)} \frac{\sin (100) \wedge (010)}{\sin (001) \wedge (010)}.$$

Monoclinic: The axial ratio is $a:1:c$. Symmetry requires $\alpha = \gamma = 90°$ $\neq \beta$.

$$\beta = 180° - (\measuredangle 5 + \measuredangle 6) = 180° - (100) \wedge (001). \qquad (19)$$

In monoclinic crystals

$$\measuredangle 3 + \measuredangle 4 = \measuredangle 1 + \measuredangle 2 = 90°;$$

therefore

$$\frac{a}{b} = \frac{\sin \measuredangle 1}{\sin \measuredangle 2} = \tan \measuredangle 1. \qquad (20)$$

Since $\measuredangle 3 + \measuredangle 4 = 90°$, we have

$$\cos \measuredangle 1 = \frac{\tan (001) \wedge (100)}{\tan (001) \wedge (110)} \quad \text{and} \quad \sin \measuredangle 1 = \frac{\sin (100) \wedge (110)}{\sin (001) \wedge (110)}. \qquad (21)$$

We also have

$$\frac{c}{b} = \frac{\sin \measuredangle 3}{\sin \measuredangle 4} = \tan \measuredangle 3. \qquad (22)$$

In a way similar, since $\measuredangle 1 + \measuredangle 2 = 90°$, we have

$$\cos \measuredangle 3 = \frac{\tan (001) \wedge (100)}{\tan (001) \wedge (111)} \quad \text{and} \quad \sin \measuredangle 3 = \frac{\sin (001) \wedge (011)}{\sin (001) \wedge (111)}. \qquad (23)$$

Finally

$$\frac{c}{a} = \frac{\sin \measuredangle 5}{\sin \measuredangle 6} = \frac{\sin (001) \wedge (101)}{\sin (100) \wedge (101)}. \qquad (24)$$

Orthorhombic: The axial ratio is $a:1:c$. Symmetry requires $\alpha = \beta = \gamma = 90°$. Because

$$\measuredangle 1 + \measuredangle 2 = \measuredangle 3 + \measuredangle 4 = \measuredangle 5 + \measuredangle 6 = 90°,$$

there exist the parametric relations

$$\frac{a}{b} = \tan \measuredangle 1 = \tan (100) \wedge (110) = \frac{h}{k} \tan (100) \wedge (hk0),$$

$$\frac{c}{b} = \tan \measuredangle 3 = \tan (001) \wedge (011) = \frac{l}{k} \tan (001) \wedge (0kl), \qquad (25)$$

$$\frac{c}{a} = \tan \measuredangle 5 = \tan (011) \wedge (101) = \frac{l}{h} \tan (001) \wedge (h0l).$$

The following also holds.

$$\frac{a}{h}\cos(100) \wedge (hkl) = \frac{b}{k}\cos(010) \wedge (hkl) = \frac{c}{l}\cos(001) \wedge (hkl). \quad (26)$$

Tetragonal: The parametric ratio is $1:1:c$. Symmetry requires $\alpha = \beta = \gamma = 90°$. The following relations hold.

$$\angle 1 + \angle 2 = \angle 3 + \angle 4 = \angle 5 + \angle 6 = 90°$$

and

$$\angle 1 = \angle 2 = 45°; \quad \angle 3 = \angle 5 \neq \angle 4 = \angle 6.$$

From these conditions, $c/a = c/b$ can be calculated in the following way.

$$\frac{c}{a} = \tan(001) \wedge (101)$$

$$= \tan(001) \wedge (011) \quad (27)$$

$$= \cos 45° \tan(001) \wedge (111).$$

Hexagonal: The parametric ratio is $1:1:c$. Symmetry requires $90° = \alpha = \beta \neq \gamma = 120°$. The following relations can be easily deduced.

$$\frac{c}{a} = \tan(0001) \wedge (10\bar{1}1)\cos 30°$$

$$= \tfrac{1}{2}\tan(0001) \wedge (11\bar{2}1) \quad (28)$$

$$= \tan(0001) \wedge (11\bar{2}2).$$

If the lattice is rhombohedral, the rhombohedral translations can be taken as reference axes. In that case, symmetry requires the axial ratio to be

$$a = b = c = 1$$

and also requires

$$\alpha = \beta = \gamma \neq 90°.$$

The calculation of the axial ratio can be made using the hexagonal reference axes, as before. We have also

$$\cos\frac{\alpha}{2} = \frac{\cos 60°}{\cos(rr/2)} \quad (29)$$

where rr is the angle between two poles of the axial rhombohedron.

Cubic: The parametric ratio is $1:1:1$. Symmetry requires $\alpha = \beta = \gamma = 90°$. The following conditions exist between the auxiliary angles.

$$\angle 1 = \angle 2 = \angle 3 = \angle 4 = \angle 5 = \angle 6 = 45°.$$

No axial ratio has to be calculated, because it is obvious.

Axial ratio and the reciprocal lattice

The calculation of the axial ratio can be simplified by using the elements a^*, b^*, c^*, α^*, β^*, γ^* of the reciprocal lattice. As in the case of the crystal lattice, goniometry enables us to deduce the ratios of a^*, b^*, and c^*, but not their absolute lengths. We can define, then, an axial ratio of the reciprocal lattice, such as

$$\frac{a^*}{b^*} : 1 : \frac{c^*}{b^*}, \tag{30}$$

corresponding to the axial ratio of the crystal lattice. The quotients of the ratio (30) can be obtained directly from angles between important poles. For instance,

$$\frac{a^*}{b^*} = \frac{\sin (010) \wedge (110)}{\sin (100) \wedge (110)},$$

$$\frac{c^*}{b^*} = \frac{\sin (010) \wedge (011)}{\sin (001) \wedge (011)}, \tag{31}$$

$$\frac{a^*}{c^*} = \frac{\sin (001) \wedge (101)}{\sin (100) \wedge (101)},$$

The ratios can also be obtained from the angles between the poles of the axial planes and the poles of the planes $(hk0)$, $(0kl)$ and $(h0l)$.

$$\frac{ha^*}{kb^*} = \frac{\sin (010) \wedge (hk0)}{\sin (100) \wedge (hk0)},$$

$$\frac{kb^*}{lc^*} = \frac{\sin (001) \wedge (0kl)}{\sin (010) \wedge (0kl)}, \tag{32}$$

$$\frac{ha^*}{lc^*} = \frac{\sin (001) \wedge (h0l)}{\sin (100) \wedge (h0l)}.$$

The usual axial ratio can then be calculated in a simple way as follows.

$$\frac{a}{b} = \frac{b^* \sin \alpha^*}{a^* \sin \beta^*}; \quad \frac{c}{b} = \frac{b^* \sin \gamma^*}{c^* \sin \beta^*}.$$

In the case of a hexagonal crystal with Miller–Bravais indices (with four integers), the calculation is the same as before:

$$\frac{a^*}{b^*} = \frac{\sin (01\bar{1}0) \wedge (11\bar{2}0)}{\sin (10\bar{1}0) \wedge (11\bar{2}0)} = \frac{\sin 30°}{\sin 30°} \,,$$

$$\frac{a^*}{c^*} = \frac{\sin (0001) \wedge (10\bar{1}1)}{\sin (1010) \wedge (10\bar{1}1)} = \tan (0001) \wedge (10\bar{1}1), \qquad (33)$$

$$\frac{b^*}{c^*} = \frac{\sin (0001) \wedge (01\bar{1}1)}{\sin (01\bar{1}0) \wedge (01\bar{1}1)} = \tan (0001) \wedge (01\bar{1}1).$$

Since $\tan (0001) \wedge (10\bar{1}1) = \tan (0001) \wedge (01\bar{1}1)$, it follows that

$$\frac{a^*}{c^*} = \frac{b^*}{c^*} \qquad (34)$$

and that

$$\frac{c}{b} = \frac{3}{2} \frac{a^*}{c^*}. \qquad (35)$$

The axial ratio can then be calculated by measuring the angles between the required poles corresponding to the Laue spots. The Laue photograph, therefore, can be used as a substitute for optical goniometry in crystallographic studies.

Classical treatment of the crystal systems

The foregoing discussion of the assignment of crystals to the crystal systems and the data derivable from the measurements has been directed especially to those who will use the Laue method for this purpose. The general subject, however, is a classical one in crystallography, and had already been well developed when goniometric measurements on crystals were made with an optical goniometer. A good introduction to the literature is given by a series of papers by Palache[9–15].

Significant literature

1. W. H. Miller [übersetzt und erweitert von J. Grailich] Lehrbuch der Krystallographie (Vienna, 1956) §4.
2. Gustav Junghann. Ein einfaches Gesetz für die Entwickelung und die Gruppirung der Krystallzonen. *Ann. der Phys.* **152** (1874) 68–95.

3. V. Goldschmidt. Ueber Entwickelung der Krystallformen. *Z. Kristallogr.* **28** (1897) 1–35, 414–451.

4. E. von Fedorow. Beiträge zur zonen Krystallographie. 1. Ein besonderer Gang der zonalen Formentwickelung. *Z. Kristallogr.* **32** (1900) 446–492.

5. E. von Fedorow. Beiträge zur zonalen Krystallographie. V. Complicationsgesetze und richtige Aufstellung der Krystalle. *Z. Kristallogr.* **35** (1902) 25–74.

6. H. Baumhauer. Untersuchungen über die Entwickelung der Krystallflächen im Zonenverbande. *Z. Kristallogr.* **38** (1904) 628–655.

7. H. Baumhauer. Die neue Entwickelung der Kristallographie. (Friedrich Vieweg, Braunschweig, 1905) 184 pages, especially 129–143 [includes a short summary of zone-theory history].

8. H. Baumhauer. Geometrische Kristallographie. Über das Gesetz der Komplikation und die Entwicklung der Kristallflächen in flächenreichen Zonen. *Fort. Mineralogie* **1** (1911) 22–37.

9. Charles Palache. Calculations in the isometric system. *Amer. Mineral.* **5** (1920) 44–48.

10. Charles Palache. Calculations in the tetragonal system. *Amer. Mineral.* **5** (1920) 49–51.

11. Charles Palache. Calculations in the hexagonal system. *Amer. Mineral.* **5** (1920) 53–59.

12. Charles Palache. Calculations in the orthorhombic system. *Amer. Mineral.* **5** (1920) 61–62.

13. Charles Palache. Calculations in the monoclinic system. *Amer. Mineral.* **5** (1920) 69–77.

14. Charles Palache. Introduction to the triclinic system. *Amer. Mineral.* **5** (1920) 77–82.

15. Charles Palache. Calculations in the triclinic system. *Amer. Mineral.* **5** (1920) 82–95.

16. Austin F. Rogers. The addition and subtraction rule in geometrical crystallography. *Amer. Mineral.* **11** (1926) 303–315.

17. M. A. Peacock. Calaverite and the law of complication. *Amer. Mineral.* **17** (1932) 317–337.

18. M. J. Buerger. The law of complication. *Amer. Mineral.* **21** (1936) 702–714.

Chapter 9

The polychromatic component

Crystal goniometry

If a crystal grows unhindered in an environment of spherical symmetry, it normally develops a set of external plane faces; the symmetry of the polyhedron composed of these faces displays the point-group symmetry of the crystal. In particular, the relative sizes of the several plane faces of the polyhedron ideally conform to the crystal symmetry. But such an ideal development requires physical conditions that are not usually met. While an unsymmetrical environment seriously affects the relative size of symmetrically equivalent crystal faces, it does not affect their orientations. If only the symmetry of the orientations of the faces of the polyhedron is considered, this corresponds to the symmetry of the lattice. There are seven such symmetries, one corresponding to the holohedral class of each crystal system except the hexagonal, to which there correspond two symmetries.

In 1809 Wollaston invented the reflecting goniometer; this simple instrument allowed the angles between the normals to crystal faces to be measured accurately for the first time. Later studies of the orientations of crystal faces were rendered even easier by the introduction of the two-circle goniometer, which was invented independently by Miller in 1874, Fedorov in 1889, and Victor Goldschmidt in 1893. These instruments make it a simple matter to measure the latitude and longitude of the position where the normal to each crystal face would reach a sphere centered on the crystal. In this way the symmetry of the orientation of the crystal faces is revealed in the orientations of their normals. As noted earlier, such a study ordinarily reveals the crystal system to which the crystal should be referred. But if the crystal belongs to a nonholohedral class of its crystal system, this lesser symmetry may not be obvious because, even in holohedral crystals, an unequal development of faces that should be equally developed is often encountered due to an unsymmetrical environment during growth.

The spherical projection

In crystal goniometry the angles between the normals to crystal faces are measured. The essence of goniometry, therefore, is the substitution of a set of normals for a set of crystal faces. A simple way to treat these normals is to locate their points of intersection with a sphere centered on the crystal. The point of intersection of a normal to a crystal face with the sphere is called the *pole* of the crystal face; the sphere is called the *pole sphere*, and the collection of poles on the sphere is called the *spherical projection* of the set of crystal faces. The radius of the sphere is taken as unity.

The position of a pole on the sphere can be defined by two coordinates which might be latitude and longitude. Actually, spherical coordinates, in the form of the colatitude ρ and the longitude ϕ, are customarily employed in crystal goniometry. Curves of ρ = constant and ϕ = constant can be drawn on the sphere. For a hemisphere, these coordinates are limited to

$$0° \leq \phi \leq 360°, \qquad 0° \leq \rho \leq 90°.$$

The actual values of the spherical coordinates ρ and ϕ of a given pole P depend obviously on the origins selected for ρ and ϕ. These origins depend on the relation between the camera axis and the direction of the x-ray beam.

The reflection sphere

In order to interpret Laue photographs it is convenient to think of the reflections as plotted on the surface of a sphere that might be called the *reflection sphere*. If the sphere is thought of as surrounding the crystal responsible for diffraction, then each x-ray beam diffracted by the crystal is a radius of the sphere (Fig. 1). The points of intersection of these reflection radii and the sphere's surface constitute a representation of the goniometry of the distribution of the x-ray reflections. Figure 1 shows that the original x-ray beam is a diameter of the reflection sphere and that the Laue reflection R_{hkl} deviates from this direction by 2θ, where θ is the Bragg

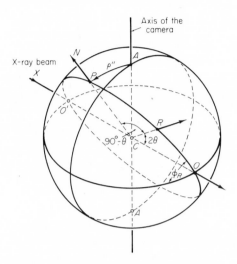

Fig. 1. Relation between pole P and reflection R.

glancing angle

$$\theta = \sin^{-1}\left(\frac{n\lambda}{2d_{hkl}}\right). \qquad (1)$$

The plane of the angle 2θ contains not only the original and reflected x-ray beam, but the normal to the reflecting plane. The Laue reflection R_{hkl} is assigned angular coordinates

$$2\theta_{hkl}, \quad \phi_{hkl},$$

where ϕ_{hkl} is the azimuth of the plane (represented by ϕ_R in Fig. 1) about the x-ray beam.

Coordinate settings

The relationship between angular coordinates ρ, ϕ of the poles P_{hkl} on the pole sphere and the angular coordinates 2θ, ϕ of the corresponding reflections R_{hkl} on the reflection sphere obviously depends on the relative orientations of the coordinate systems of the two reference spheres. It may be convenient to choose different coordinate systems depending on the setting of a certain symmetry direction in the crystal with respect to the x-ray beam. In every case both spheres are supposed to have the same center, the center of the crystal. In principle it is possible to consider any kind of setting, and Fig. 2 gives the relative orientations that define certain

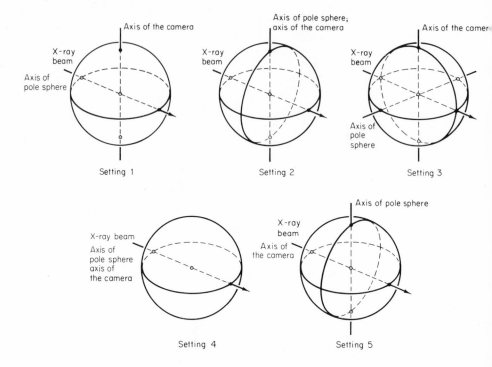

Fig. 2. Settings derived from the different relative orientations of the pole sphere and the axis of the camera.

different settings[5]. The setting becomes very important when a cylindrical camera is used to record the Laue reflections. In this case, it is necessary to define a third axis as well, the camera axis, which in flat-film techniques can be taken as the direction along which the crystal is mounted. In order to simplify the nomenclature we shall also call this direction the camera axis. It is obvious that the camera axis is not necessarily a rational crystallographic direction.

The possible special settings are the following:

Settings 1, 2, 3: Camera axis perpendicular to x-ray beam;
    Setting 1: axis of pole sphere along x-ray beam;
    Setting 2: axis of pole sphere along camera axis;
    Setting 3: axis of pole sphere perpendicular to both x-ray beam and
          camera axis.

Any other setting in which the camera axis and the x-ray beam are mutually perpendicular is of no practical importance. There exist, however, cameras in which the x-ray beam coincides with the camera axis. For this particular case two more settings can be given:

Settings 4, 5: Camera axis along x-ray beam;
  Setting 4: axis of pole sphere along x-ray beam;
  Setting 5: axis of pole sphere perpendicular to x-ray beam.

Settings 4 and 5 are used only in very special cases. The most widely used Laue photographs are those taken with the x-ray beam and camera axis mutually perpendicular; the first three settings then have practical importance. The most obvious relations exist between the pole coordinates and the reflection coordinates in setting 1, and this setting will be used throughout this monograph. However, setting 2 also has some merit, especially because it has been widely used, and it will be thoroughly discussed in this chapter. For the sake of completeness, the other settings, although of less practical importance, will also be discussed.

Relation between coordinates in the pole sphere and in the reflection sphere for setting 1

Let us make the pole sphere coincide with the reflection sphere, and the axis of the pole sphere with the x-ray beam axis (Fig. 3i). In addition, let the origins of the pole sphere and reflection sphere be at opposite ends of the same diameter, so that the origin of the polar coordinate ρ' is at point O' (the intersection of the x-ray beam at the entrance side of the reflecting sphere). The origin of the meridian ϕ' is taken in the horizontal plane containing the x-ray beam and perpendicular to the camera axis, ϕ' increasing in the counterclockwise sense when viewed in the $O'O$ direction from the crystal.

According to Descartes' law, the incident x-ray beam $O'O$, the normal CP to the crystal plane, and the reflected ray CR lie in the same plane. Hence, P_{hkl} and R_{hkl} lie on the same meridian (Fig. 3ii), and we can write

$$\phi'_{P_{hkl}} = \phi_{R_{hkl}}. \tag{2}$$

It follows that the set of curves $\phi_{P}' = $ constant on the pole sphere transforms into the same set of curves $\phi_{R} = $ constant on the reflection sphere; these are half great circles (meridians) measuring from 0° to 360° in the counterclockwise direction as seen looking along the direction $O'O$.

On the other hand, Fig. 3ii shows that the symmetry about CP requires

$$2\rho' + 2\theta = 180°; \tag{3}$$

hence,

$$2\theta_{hkl} = 2(90° - \rho'_{hkl}). \tag{4}$$

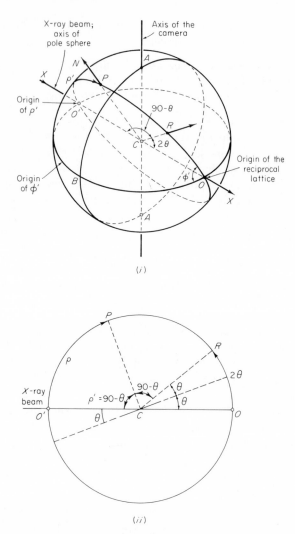

Fig. 3. Relations between pole P and Laue reflection R in setting 1. (i) O, origin of the reciprocal lattice; O', origin of ρ'; OBO', origin of ϕ', (ii) Great circle common to the pole P and Laue reflection R.

From (4) it follows that the curves of ρ' = constant, which are parallels on the pole sphere, are easily transformed into 2θ curves, which are also parallels in the reflection sphere.

Setting 1 has the great advantage that meridians and parallels are not only the loci of ϕ_P' = constant and ρ_R' = constant on the pole sphere, but are also the loci of $\phi_R' = $ constant and $2\theta = $ constant on the reflection sphere. A set of meridians and parallels corresponding to curves of 2θ and ϕ' at constant intervals can be plotted on the reflection sphere. If the set of 2θ curves is labeled in terms of the corresponding ρ' deduced from (4), the spherical coordinates ρ'_{hkl} and ϕ'_{hkl} of the pole of a stack of planes (hkl) that has given rise to a Laue reflection R_{hkl} on the reflection sphere could be determined directly from the position of R_{hkl}. In order to make use of this feature, let us further analyze the relations between the sets of ρ' and 2θ curves.

The Laue reflections extend all over the reflection sphere. However, the poles that generated such reflections are concentrated in half the polar sphere. In fact, a plane with $\rho' = 90°$ gives a reflection with $\theta = 0$, that is, in the direction of the primary beam. A pole with $\rho' = 0$ corresponds to a stack of planes perpendicular to the x-ray beam and gives a reflection with $\theta = 90°$, which is in a direction opposite to that of the incident x-ray beam. The point where the x-ray beam enters the reflection sphere corresponds to a pole of $\rho' = 0$ and the point where the x-ray beam emerges from the same sphere corresponds to a pole of $\rho' = 90°$. By labeling the parallels of the reflection sphere at steps of $2°$ of $2\theta = 1°$ of ρ', starting from the point where the x-ray beam enters the sphere, the curves of 2θ give a direct reading of ρ'.

Transformation equations between film and pole coordinates in setting 1

In principle the diffracted rays coming from a stationary crystal could be recorded on a spherical film but this would be an impractical kind of film. A practical procedure is to record them on a cylindrical film whose axis is parallel to the rotation axis of the camera, or else on a plane film placed normal to the x-ray beam, in either back or front position. When this is done, the pattern of Laue spots on the film is not the same as the pattern of reflecting points on the sphere, but the former is a projection of the latter. Thus, every Laue spot with coordinates x, y on the film corresponds with a unique ρ', ϕ' coordinate pair. Transformation from film coordinates x, y to sphere pole coordinates ρ', ϕ' can be obtained through the intermediate step of the 2θ, ϕ' coordinates on the reflection sphere. It is then possible to seek the pole P_{hkl} that corresponds with the Laue spot L_{hkl} on the film and thus to prepare a projection of the Laue photograph from which it can be indexed.

Two different approaches can be used to solve this problem. In the first method, the film coordinates x, y of each Laue spot are measured; by applying the appropriate transformations, the corresponding pole coordinates ρ', ϕ' of the stack of planes giving rise to the Laue spots are found. In the second method a chart is prepared which maps the ρ', ϕ' pole coordinates on the film. This allows a direct reading of the pole coordinates ρ', ϕ' of each plane that produces a Laue spot on the film. This second procedure is the most commonly used. However, the first approach, which has been avoided for many years, is becoming more appealing today due to the ready availability of high-speed computers[23].

The relations between the pole coordinates of the Laue reflection and the film coordinates can be deduced in a straightforward way using an auxiliary system of pole coordinates independent of the kind of setting used. In this system the position of a Laue reflection R on the reflection sphere is defined in terms of the spherical coordinates Υ (upsilon), called the azimuth angle, which is normal to the plane containing the reflection and the camera axis, and χ (chi), the inclination angle, which is in the plane containing R and the camera axis AA' (Fig. 4). The angles χ and Υ, and 2θ and ϕ' are related by the spherical trigonometric relations

$$\cos 2\theta = \cos \chi \cos \Upsilon, \tag{5}$$

$$\tan \phi' = \tan \chi \operatorname{cosec} \Upsilon. \tag{6}$$

The ρ' and ϕ' pole coordinates of every recorded Laue spot can easily be derived in terms of χ and Υ from measurements made on the cylindrical or flat film.

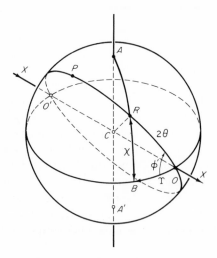

Fig. 4. Definition of the auxiliary angles χ and Υ.

Cylindrical Laue photographs

Ordinarily a Laue photograph taken with a cylindrical camera employs an incident x-ray beam perpendicular to the axis of the cylinder. These conditions are represented in Fig. 5. In this case, the angle Υ is related to the film coordinate x by the arc–angle proportion

$$\frac{\Upsilon}{360°} = \frac{x}{2\pi r_F} \tag{7}$$

where r_F is the radius of the film. From this

$$\Upsilon = \frac{x}{r_F}\left(\frac{360°}{2\pi}\right). \tag{8}$$

The angle χ is determined from the film coordinate y by

$$\tan \chi = \frac{y}{r_F}. \tag{9}$$

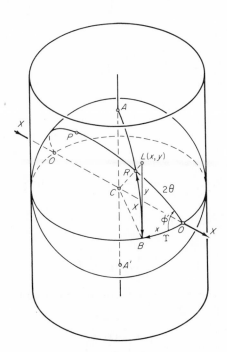

Fig. 5. Relations between cylindrical-film coordinates and spherical coordinates.

From (4) it is evident that

$$\cos 2\theta = -\cos 2\rho'. \tag{10}$$

Substituting in (5) the value of Υ given by (8) and the value of χ given by (9) yields

$$\cos 2\rho' = -\cos\left[\frac{x}{r_F}\left(\frac{360°}{2\pi}\right)\right]\cos\left[\tan^{-1}\left(\frac{y}{r_F}\right)\right]. \tag{11}$$

In a similar way, substituting in (6) the values of (8) and (9) we have

$$\tan \phi' = \frac{y}{r_F}\cosec\left[\frac{x}{r_F}\left(\frac{360°}{2\pi}\right)\right]. \tag{12}$$

Thus, the appropriate transformations for calculating the pole coordinates ρ' and ϕ' of any pole P corresponding to a Laue spot are given in terms of the x and y coordinates measured on the cylindrical film by

$$\rho' = 90° - \tfrac{1}{2}\cos^{-1}\left\{-\cos\left[\frac{x}{r_F}\left(\frac{360°}{2\pi}\right)\right]\middle|\cos\left[\tan^{-1}\left(\frac{y}{r_F}\right)\right]\middle|\right\} \tag{13}$$

and

$$\phi' = \tan^{-1}\left\{\frac{y}{r_F}\cosec\left[\frac{x}{r_F}\left(\frac{360°}{2\pi}\right)\right]\right\}. \tag{14}$$

The film coordinates x, y of a Laue reflection can now be calculated from the spherical coordinates ρ', ϕ' of the pole corresponding to the reflection R. From (11) the film coordinate x can be computed for given values of y in terms of ρ', thus,

$$x = r_F\left(\frac{2\pi}{360°}\right)\cos^{-1}\left\{-\cos 2\rho'\sec\left[\tan^{-1}\left(\frac{y}{r_F}\right)\right]\right\}, \tag{15}$$

and in terms of ϕ from (12):

$$x = r_F\left(\frac{2\pi}{360°}\right)\sin^{-1}\left(\frac{y}{r_F \tan \phi'}\right). \tag{16}$$

On the other hand, the coordinate y can be computed for given values of x in terms of ϕ' through (12):

$$y = r_F \tan \phi' \sin\left[\frac{x}{r_F}\left(\frac{360°}{2\pi}\right)\right], \tag{17}$$

and in terms of ρ' from (11):

$$y = r_{\mathrm{F}} \tan\left(\cos^{-1}\left\{\cos 2\rho' \sec\left[\frac{x}{r_{\mathrm{F}}}\left(\frac{360°}{2\pi}\right)\right]\right\}\right). \tag{18}$$

Equations (15), (17) and (16), (18) give the position where a Laue spot is recorded on a cylindrical film corresponding to a pole of known ρ' and ϕ' coordinates.

Flat Laue photographs

When a flat film is used to record the Laue photograph, the film is usually perpendicular to the x-ray beam. Two possibilities are normally used: in one case (Fig. 6i), the film is located beyond the crystal (*transmission* case), and in the other case (Fig. 6ii) it is located before the crystal (*back-reflection* case). The relations between film and Laue-reflection coordinates are somewhat different and have to be considered independently.

Transmission case. Suppose the film is tangent to the reflection sphere at the point of emergence of the x-ray beam (Fig. 6i). The ρ' and ϕ' coordinates of the pole associated with the stack of planes giving rise to a Laue spot on a transmission flat Laue photograph may be expressed in terms of the x, y film coordinates by the corresponding relations

$$\tan 2\theta = \frac{s}{D}, \tag{19}$$

D being the crystal-to-plate distance and s the distance of the Laue spot from the center (record of the primary beam) of the photograph.

Relation (19) may also be expressed directly in terms of film coordinates x and y by noting that

$$s^2 = x^2 + y^2. \tag{20}$$

From (4) it is evident that

$$\tan 2\theta = -\tan 2\rho'. \tag{21}$$

Making this substitution in (19), and by using (20), we have

$$\tan 2\rho' = -\frac{(x^2 + y^2)^{1/2}}{D}. \tag{22}$$

On the other hand, the ϕ' coordinate is easily obtained in terms of film coordinates x and y by the relation

$$\tan \phi' = \frac{y}{x}. \tag{23}$$

The desired formulas are

$$\rho' = 90° - \tfrac{1}{2}\tan^{-1}\left[-\frac{(x^2 + y^2)^{1/2}}{D}\right], \tag{24}$$

and

$$\phi' = \tan^{-1}\left(\frac{y}{x}\right). \tag{25}$$

The film coordinates x and y can be easily expressed in terms of ρ' and ϕ' from (19), (20), and (23) as follows.

$$x = \pm(D^2 \tan^2 2\rho' - y^2)^{1/2} \tag{26}$$

and

$$y = x \tan \phi'. \tag{27}$$

Back-reflection case. The planes responsible for the Laue spots in the back-reflection region are almost normal to the incident beam. In this case, the reflected ray makes an angle of $180° - 2\theta$ with the incident beam.

From Fig. 6*ii* it follows that

$$\tan(180° - 2\theta) = \frac{s}{D}, \tag{28}$$

$$\tan 2\rho' = \frac{(x^2 + y^2)^{1/2}}{D}. \tag{29}$$

Substituting (4) and (20) in this expression, we have

$$\rho' = \tfrac{1}{2}\tan^{-1}\left[\frac{(x^2 + y^2)^{1/2}}{D}\right]. \tag{30}$$

On the other hand,

$$\tan \phi' = -\frac{y}{x}; \tag{31}$$

consequently,

$$\phi' = \tan^{-1}\left(-\frac{y}{x}\right). \tag{32}$$

In this case the necessary formulas are (30) and (32).

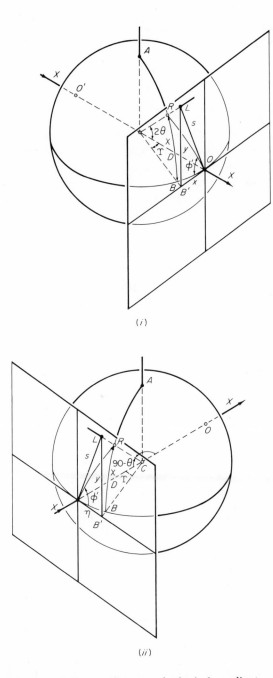

Fig. 6. Relations between film coordinates and spherical coordinates. (*i*) Transmission case; (*ii*) back-reflection case.

The transformation from polar coordinates to film coordinates can be deduced from (28) and (29):

$$x = \pm(D^2 \tan^2 2\rho' - y^2)^{1/2};\tag{33}$$

and from (25):

$$y = -x \tan \phi'.\tag{34}$$

The several transformations for deriving pole coordinates and film coordinates from one another for the first setting are assembled in Table 1.

A relationship exists between the x, y coordinates of a Laue spot recorded on a flat film and those recorded on a cylindrical film. If for the sake of clearness we use subindices c and t to represent cylindrical or transmission coordinates, respectively, we get from Fig. 7

$$x_t = r_F \tan \Upsilon,\tag{35}$$

and by using (8)

$$x_t = r_F \tan\left[\frac{x_c}{r_F}\left(\frac{360°}{2\pi}\right)\right].\tag{36}$$

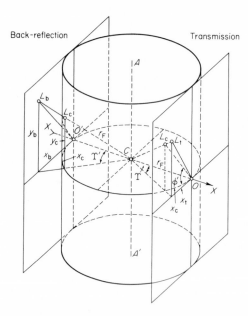

Fig. 7. Relations between the x, y coordinates of a cylindrical and a flat film in the transmission and back-reflection cases.

Table 1

Transformations for deriving pole coordinates ρ', ϕ' and film coordinates x, y from one another in setting 1

Cylindrical film	Flat film	
	Transmission case	Back-reflection case
$\rho' = 90° - \frac{1}{2}\cos^{-1}\left\{-\cos\left[\dfrac{x}{r_{\mathrm{F}}}\left(\dfrac{360°}{2\pi}\right)\right]\cos\left[\tan^{-1}\left(\dfrac{y}{r_{\mathrm{F}}}\right)\right]\right\}$ (13)	$90° - \frac{1}{2}\tan^{-1}\left[-\dfrac{(x^2+y^2)^{1/2}}{D}\right]$ (24)	$\frac{1}{2}\tan^{-1}\left[\dfrac{(x^2+y^2)^{1/2}}{D}\right]$ (30)
$\phi' = \tan^{-1}\left\{\dfrac{y}{r_{\mathrm{F}}}\cosec\left[\dfrac{x}{r_{\mathrm{F}}}\left(\dfrac{360°}{2\pi}\right)\right]\right\}$ (14)	$\tan^{-1}\left(\dfrac{y}{x}\right)$ (25)	$\tan^{-1}\left(-\dfrac{y}{x}\right)$ (32)
$x = r_{\mathrm{F}}\left(\dfrac{2\pi}{360°}\right)\cos^{-1}\left\{-\cos 2\rho'\sec\left[\tan^{-1}\left(\dfrac{y}{r_{\mathrm{F}}}\right)\right]\right\}$ (15)	$\pm(D^2\tan^2 2\rho' - y^2)^{1/2}$ (26)	$\pm(D^2\tan^2 2\rho' - y^2)^{1/2}$ (33)
$y = r_{\mathrm{F}}\tan\phi'\sin\left[\dfrac{x}{r_{\mathrm{F}}}\left(\dfrac{360°}{2\pi}\right)\right]$ (17)	$x\tan\phi'$ (27)	$-x\tan\phi'$ (34)

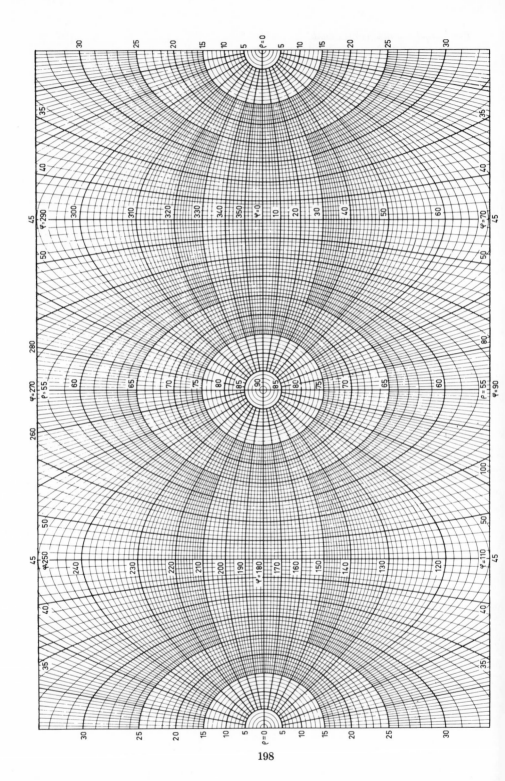

The value of x_c must satisfy

$$|x_c| < \frac{\pi}{2} r_F. \tag{37}$$

On the other hand, also from Fig. 7,

$$\frac{y_t}{y_c} = \frac{(r_F^2 + x_t^2)^{1/2}}{r_F}, \tag{38}$$

from which

$$y_t = \frac{(r_F^2 + x_t^2)^{1/2}}{r_F} y_c. \tag{39}$$

For back-reflection case, if we call x_b, y_b the film coordinates, we have from Fig. 7

$$x_b = r_F \tan \Upsilon' \tag{40}$$

where

$$\Upsilon' = \frac{\pi r_F - x_c}{r_F} \left(\frac{360°}{2\pi}\right) \tag{41}$$

and $|x_c|$ satisfies

$$|x_c| > \frac{\pi}{2} r_F. \tag{42}$$

Therefore

$$x_b = -r_F \tan\left[\frac{\pi r_F - |x_c|}{r_F} \left(\frac{360°}{2\pi}\right)\right]. \tag{43}$$

The sign of x_b is opposite to the sign of x_c.

The y coordinate is given by (39) as in the transmission case, that is,

$$y_b = y_t. \tag{44}$$

Charts for the determination of the ϱ', ϕ' coordinates directly from the film in setting 1. A very practical way of interpreting Laue photographs is to prepare a chart of the film on which the ρ', ϕ' coordinates of each Laue spot can be read directly. A chart of this kind can be easily prepared once the transformations from polar to film coordinates are known

Fig. 8. Chart giving ρ', ϕ' for cylindrical Laue photographs taken with setting 1. Diameter of camera is 57.3 mm. [After Canut and Amoros,[18] p. 20.]

for different kinds of film setting. The corresponding transformation formulas have been given: (13) and (14) for a cylindrical film, and (24) and (25) or (30) and (32) for a flat plate. The construction of the charts is easy. It is accomplished by assuming values of the pair ρ', ϕ' at constant steps of 1°, plotting the values at the corresponding x, y film coordinates, and then connecting with a line all the points for which ρ' is equal to 1°, and so on. The same procedure can be followed concerning points with $\phi' =$ constant. The lines so drawn form a net over the whole area of the film; from this net the values of ρ' and ϕ' of a pole that corresponds to a Laue spot may be read for every Laue spot in the photograph. Charts have been computed by us for both cylindrical films and flat plates. In Fig. 8 the chart for a cylindrical camera of radius of 28.65 mm (1 mm $x = 2°$, which is a particularly convenient dimension) is reproduced.

A table of values to construct a somewhat similar chart is given in volume II of the International tables for x-ray crystallography[19]. However, that table was based on values of θ (the glancing angle) instead of ρ' (the pole coordinate).

Figure 9 shows the location of spots from a Laue photograph in which trails of spots bear a striking resemblance to the lines of constant ϕ' of the chart just described. The crystal (stearic acid, a long-chain monocarboxylic acid) was mounted with its a axis parallel to the camera axis and its c axis parallel to the x-ray beam. The crystal is characterized by a very short reciprocal-lattice spacing along the c axis. The Laue cones are very densely populated, so the Laue spots form almost continuous lines of constant ϕ'.

A chart valid for both transmission and back-reflection Laue photographs for a crystal-to-film distance of 3 cm is reproduced in Fig. 10. In order to make this chart useful for both cases the ρ' and ϕ' values for back reflection are given in italics, along with the ρ' and ϕ' values of the transmission Laue case.

In the derivation of the equations from which the charts have been constructed, the film should always be viewed from the position of the crystal in the experiment. Hence, it is necessary to identify the position of the film during its exposure to the x rays. This can be done, for instance, by making a penciled check mark in the same corner of the film before it is placed on the film holder. A normal procedure is to mark the upper right-hand corner.

In practice, it is very convenient to have the chart reproduced on a transparent plastic sheet. The x-ray film is placed over this chart with its center coinciding with the film center (record of the primary x-ray beam) and with the edges of the chart and film parallel. When these two transparencies are placed against an illuminated opal-glass background, the pole coordinates ρ', ϕ' for each Laue spot may be read very easily.

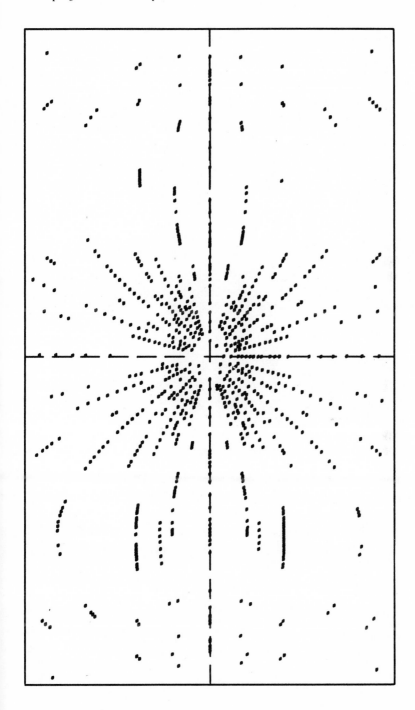

Fig. 9. Plot of spot locations from a cylindrical Laue photograph of stearic acid showing the curves of constant ϕ with setting 1. Diameter of camera is 57.3 mm.

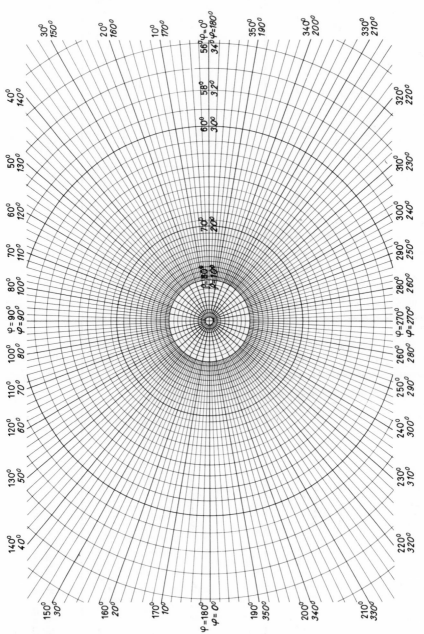

Fig. 10. Chart for ρ', ϕ' transmission and back-reflection Laue photographs made using setting 1. Crystal-to-plate distance, 3 cm. Numerals in italics are back-reflection angles. [After Canut and Amoros,[18] p. 22.]

Relation between coordinates on the pole sphere and on the reflection sphere for settings 2 and 3

In settings 2 and 3 the axis of the pole sphere is perpendicular to the x-ray beam axis. The position of the camera axis is immaterial if the relations between the pole sphere and the reflection sphere are considered alone, but the camera axis becomes important for the interpretation of the actual photographs.

In these two settings the origin of ρ and ϕ is different than in the first setting, and therefore the coordinates ρ'', ϕ'' of the pole P_{hkl} are different than the coordinates ρ' and ϕ' of the same pole in the first setting. Figure 11 gives the relations between the pole coordinates and the reflection coordinates in this new setting. The origin of ρ'' is the point A and the origin of ϕ'' is the meridian perpendicular to the incident x-ray beam and containing the axis of the pole sphere. The curves of constant ρ'' are small circles concentric with A and the curves of constant ϕ'' are meridians. In this setting the values of ρ'' are extended from $0°$ to $90°$ and from $0°$ to $-90°$ ($\rho'' = -90°$ and $\rho'' = 90°$ coincide). The possible values of ϕ'' are from $0°$ to $180°$. For every pole P_{hkl} and its reflection R_{hkl}, a geometry identical to the one discussed in Fig. 3*ii* occurs. In view of the new values ρ'', ϕ'', however, the following relation holds

$$\sin \theta = \sin \rho'' \sin \phi''. \tag{45}$$

The glancing angle is now a function of ρ'' and ϕ'', not of ρ alone, as in the

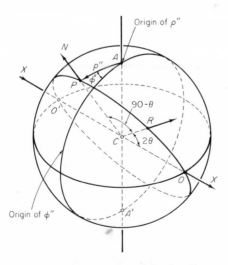

Fig. 11. Relations between pole coordinates and reflection coordinates in setting 2.

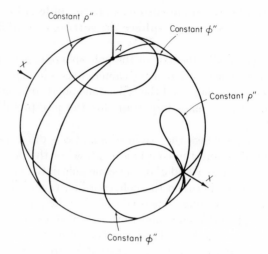

Fig. 12. Line of constant ρ'' and constant ϕ'' on the reflecting sphere.

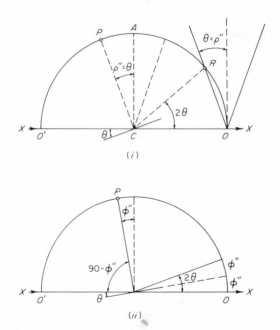

Fig. 13. Setting 2. (*i*) Plane $O'AO$ of Fig. 11 containing the axis of the pole sphere: relations between ρ'' and 2θ. (*ii*) Plane perpendicular to the axis of the pole sphere: relations between ϕ'' and 2θ.

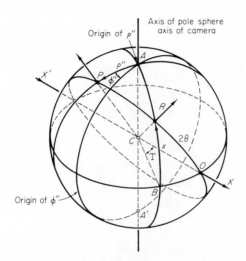

Fig. 14. Relations between ρ'', ϕ'' and Υ, χ in setting 2.

first setting. From (45) it follows that

$$\theta = \rho'' \quad \text{for} \quad \phi'' = 90°, \tag{46}$$

$$\theta = \phi'' \quad \text{for} \quad \rho'' = 90°.$$

It is evident that θ is zero when ρ'' is zero.

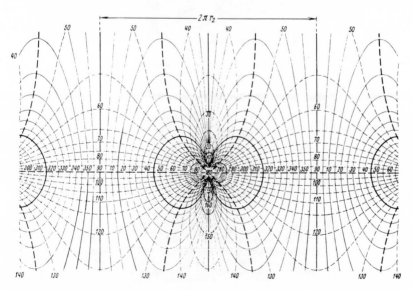

Fig. 15. Chart for ρ'', ϕ'' for cylindrical Laue photographs in setting 2. [From Rösch[5].]

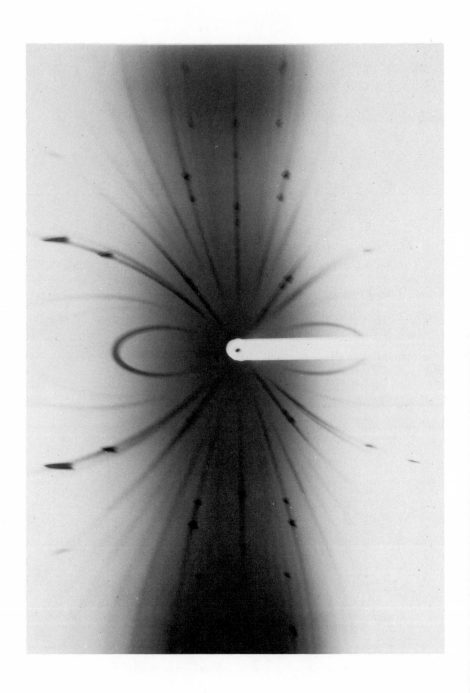

Following the scheme set forth for setting 1, it is necessary now to find out what set of curves, on the reflection sphere, corresponds to constant values of ρ'', and ϕ''. We consider first a set of poles of constant ρ''. These poles are located in a small circle around A. The reflection R is at the intersection of the reflection sphere with a cone of half angle ρ''. The axis of this cone is through the origin of reciprocal space and parallel to CA. The curves are a sort of lemniscate on the surface of the reflection sphere, Fig. 12.

Next we consider the curves of constant ϕ''. In the particular case of $\phi'' = 90°$, the relations between ρ'' and 2θ are shown in Fig. 13. It can be easily seen that the reflections corresponding to poles of constant ϕ'' are at the intersection of the sphere with a cone of half angle ϕ''. The center of the cone is C and the cone axis is perpendicular to the meridian plane ϕ''. One generator of the cone is evidently the x-ray beam. The intersection of a central cone with the sphere is a circle, and therefore the curves of constant ϕ'' on the reflection sphere are circumferences; their centers lie on the great circle perpendicular to the axis of the pole sphere at an angular distance ϕ'' from O for $\phi'' < 90°$. For $\phi'' = 90°$, the curve is the great circle containing the axis of the pole sphere and x-ray beam. For $\phi'' > 90°$ the cones have a half-opening angle equal to $180° - \phi''$ and the axis of the cone is at a distance $180° - \phi''$ from O. The zone axis of any meridian ϕ'' forms the same angle ϕ'' with the incident beam. The centers of the circles of constant ϕ'' are obviously the poles of the zone axis of the meridians.

Transformation equations between the film and pole coordinates in setting 2

The curves for constant ρ'' and constant ϕ'' project on the film in different ways in settings 2 and 3. In setting 2, the camera axis coincides with the axis of the pole sphere. In this case, the azimuth angle Υ is equal to ϕ''. The following relation can be deduced by using Fig. 14.

$$\cos 2\theta = \cos \phi'' \cos \chi; \tag{47}$$

$$\sin \chi = 2 \sin \theta \sin \rho''. \tag{48}$$

Taking relations (47) and (48) into account, it is easy to calculate the lines of constant ρ'' and constant ϕ'' as projected on the cylindrical or flat film, which is dealt with in the following sections.

Fig. 16. Rotation diagram showing the ρ'' curves of setting 2. [From M. J. Buerger, X-ray crystallography, (Wiley, New York, 1942) p. 153, Fig. 87].

Cylindrical Laue photographs. For a cylindrical camera, χ and ϕ'' are given directly in terms of x and y by

$$\phi'' = \frac{x}{r_F}\left(\frac{360°}{2\pi}\right) \tag{49}$$

and

$$\tan \chi = \frac{y}{r_F}. \tag{50}$$

Combining (45) and (46) with the previous two equations, we have

$$\cos 2\theta = \cos\left[\frac{x}{r_F}\left(\frac{360°}{2\pi}\right)\right]\cos \tan^{-1}\left(\frac{y}{r_F}\right) \tag{51}$$

and

$$\sin \tan^{-1}\left(\frac{y}{r_F}\right) = 2 \sin \theta \sin \rho''. \tag{52}$$

The ρ'' and ϕ'' coordinates of a pole in the reflecting sphere are given by

$$\rho'' = \sin^{-1}\left[\frac{\sin \tan^{-1}(y/r_F)}{2 \sin \theta}\right], \tag{53}$$

$$\phi'' = \cos^{-1}\left[\frac{\cos 2\theta}{\cos \tan^{-1}(y/r_F)}\right]. \tag{54}$$

A chart for the cylindrical Laue photograph is given by Henry, Lipson and Wooster[15] corresponding to setting 2 (Fig. 15). The ϕ'' curves are similar to lemniscates when ϕ'' is less than 45°, and are sinuous when ϕ'' is greater than 45°. They are the curves along which the orders of reflection of different radiation lie for each reflecting plane. The ρ'' curves coincide with the ρ curves displayed in a rotating-crystal photograph; an example is shown in Fig. 16 for a large diamond crystal rotating about its a axis, using unfiltered Cu*K* radiation.

Flat Laue photographs. Let us consider the curves of constant ρ'' and constant ϕ'' on the reflection sphere as projected on a flat film. The lines of constant ρ'' are curves of the fourth degree for the transmission photographs. The lines of constant ϕ'' are conics, specifically, ellipses and a parabola in the transmission region, and hyperbolas in both regions.

According to Bernalte[21], the curves of constant ρ'' on the flat Laue photo-

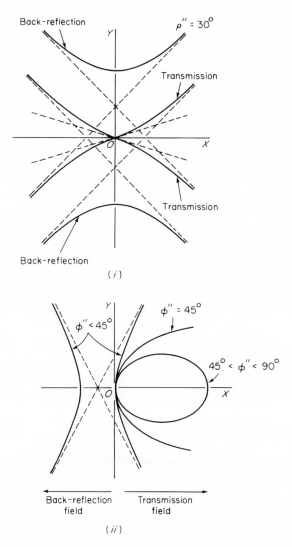

Fig. 17. (*i*) Curves of $\rho'' = 30°$ for the flat Laue transmission and back-reflection cases. (*ii*) The three types of conics in transmission and back-reflection Laue photographs. [From Bernalte[21], p. 918, modified.]

graphs are of degree four,

$$[y^2 - (\cot \rho'' \cot 2\rho')x^2]^2 + D^2[y^2 - (\cot^2 \rho'')x^2] = 0 \qquad (55)$$

with three biflectnodes. One of these occurs at the origin with tangents

$$y - (\cot \rho'')x = 0, \qquad (56)$$

while the other two occur at infinity with asymptotes

$$y \pm \cot \rho''(\cot 2\rho'')x \pm \frac{\frac{1}{2}D(\cot^2 \rho'' \cot^2 2\rho'')^{1/2}}{\cot \rho'' \cot 2\rho''} = 0. \qquad (57)$$

Figure 17i represents the curve $\rho'' = 30°$ on the flat Laue photograph for the transmission and back-reflection cases.

The curves of constant ϕ'' in the flat Laue photograph are conics:

$$x^2 - (\cot^2\phi'' - 1)y^2 + 2D(\cot \phi'')y = 0. \qquad (58)$$

These conics are hyperbolas for $\phi'' < 45°$, a parabola for $\phi'' = 45°$, and ellipses for $45° < \phi'' < 90°$.

In the transmission region, all three kinds of conics appear. But in the back-reflection Laue photographs, only the hyperbolas are shown, with branches which do not pass through the origin. Figure 17ii represents the three types of conics in transmission and back-reflection Laue photographs.

Transmission case. For this recording arrangement, a chart due to Leonhardt[4] has been widely used. In the original Leonhardt chart the constant-ρ'' quadrics are labeled from 0° to 90°, starting at the point of the x-ray beam incidence. Dunn and Martin[14], Cullity[17], Bernalte[21], and

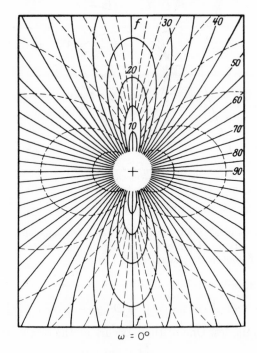

$\omega = 0°$

Fig. 18i

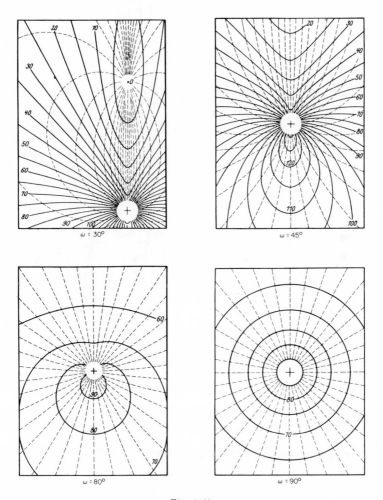

Fig. 18ii

Fig. 18. Charts for ρ'', ϕ'', in setting 2 for different positions of the flat film. Angle ω refers to the inclination of the plate normal in the plane of the crystal axis and x-ray beam. Chart for $\omega = 0$ has a crystal-to-film distance of 3 cm; for other charts this distance is 2 cm. [From Leonhardt[4], pp. 103–105, Figs. 1–5.]

Azaroff[22], however, designate these curves on the Leonhardt chart in terms of δ, where $\delta = 90° - \rho'$. For the sake of uniformity in relation to the ρ'' curves displayed in cylindrical Laue photographs, we follow Leonhardt in the numbering of the ρ'' curves. The labels of the curves of constant ϕ'' as given by Leonhardt have, however, been maintained by most authors, and since they correspond to the ones adopted in this book for the cylindrical

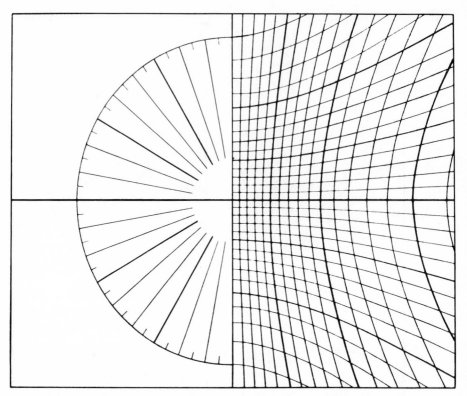

Fig. 19. Chart for ρ'', ϕ'', setting 2 for back-reflection Laue photograph, Crystal-to-film distance, 3 cm. Vertical curves are meridians; horizontal curves are parallels. Intervals are 2°. [From Greninger[12].]

case, we shall also follow Leonhardt with respect to the labeling of meridian curves. Figure 18*i* gives a Leonhardt chart for a crystal-to-film distance of 3 cm.

Leonhardt also studied the variation of the curves of constant ρ'' and constant ϕ'' for the case in which the photographic plate is rotated through an angle ω. The ρ'', ϕ'' curves for $\omega = 0°$, 30°, 45°, 80°, and 90° are represented in Fig. 18*ii*. Obviously for $\omega = 90°$ the curves coincide with the ρ', ϕ' curves for the transmission case of the first setting.

Back-reflection case. For this particular case of recording, a chart first given by Greninger[12] has been widely used. Figure 19 shows the Greninger chart for a crystal-to-film distance of 3 cm. For the sake of uniformity with the cylindrical case, the curves of constant ρ'' and constant ϕ'' on this chart are labeled following Greninger. The curves closest to the center of the photograph correspond to 90°, as in the back-reflection region of the cylindrical chart of Henry, Lipson, and Wooster[15]. Cullity[17] uses δ and λ curves; they are related to ours by $\delta = 90° - \rho''$ and $\Upsilon = 90° - \phi''$.

212

Other coordinate settings

The three remaining settings are less important. Reference can be found in the International *Tabellen*[11]. In setting 3 the camera axis is perpendicular to both the axis of the pole sphere and the x-ray beam. The curves of constant ρ''' and constant ϕ''' in the reflection sphere are identical to those of setting 2, but the new position of the camera axis makes the projection of such curves on the cylinder of the camera different than in setting 2. The lemniscates of constant ρ''' are now projected along the equatorial direction in the cylinder and the great circles of constant ϕ''' are projected along the vertical axis of the cylinder. The result is that the whole chart (Fig. 20) is not very different from the chart of setting 2. Figure 21 shows a CalComp plot of the spots on a Laue photograph in which the trail of spots bears a striking resemblance to the curves of constant ϕ''' of the chart just described. This Laue photograph is from stearic acid mounted so that the *a* axis is parallel to the camera axis and the *b* axis parallel to the x-ray beam. As noted before, the crystal is characterized by a very short reciprocal-lattice spacing along the *c* axis, so the Laue cones are densely populated, giving rise to almost continuous lines of constant ϕ'''. The special relations between the camera axis, the pole sphere, and reflection

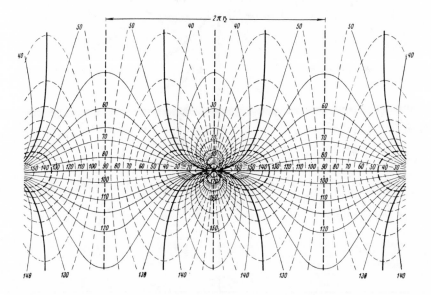

Fig. 20. Chart for ρ''', ϕ''', setting 3, for cylindrical Laue photograph. [From Schiebold and Schneider[11], p. 636, Fig. 21b.]

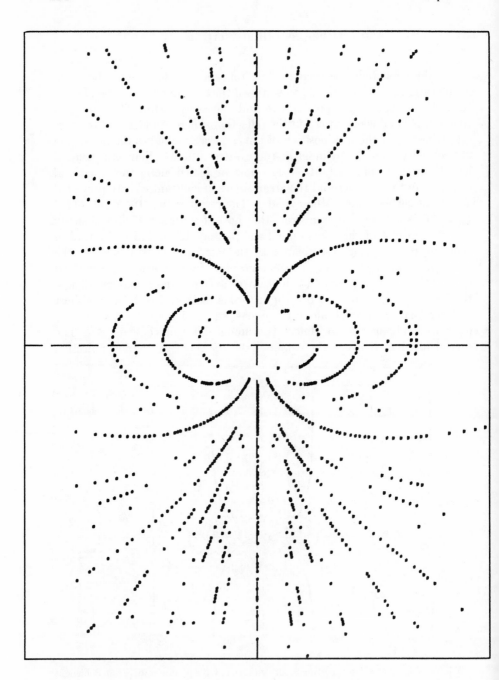

sphere determine the trigonometric expressions between the angles. For instance, in setting 3 the following relation can easily be obtained.

$$\sin \chi = \frac{\tan \theta \sin 2\theta}{\tan \phi'''} , \tag{59}$$

$$\tan \Upsilon = \frac{\tan 2\theta \tan \phi'''}{\tan \theta} , \tag{60}$$

and so on.

In the case of a flat film, the direction of the camera axis is immaterial to the projection of the curves of constant ρ''' and constant ϕ''' as well as of constant ρ'' and constant ϕ''. The charts are identical to the Greninger[12] and Leonhardt[4] charts, but have been rotated 90° for setting 3.

In setting 4 the lines of constant ρ'''' are projected as lines parallel to the equator of the cylinder; the curves of constant ϕ'''' are lines parallel to the axis of the cylinder (Fig. 22*i*). Setting 4 is especially applicable to Laue cameras in which the photographic film is in the form of a cone whose axis is parallel to the x-ray beam. Cameras of this type have been employed by Regler[7,8] and Herglotz[16,20]. When the opening angle of the cone is 60°, the film is half a circle, which is easily cut. The positions on the film at which the focusing (discussed in Chapter 13) occurs can be varied by adjusting the location of the entrance pinhole. The cone may be pointed toward the x-ray source to record the entire back-reflection field, or in the opposite direction to record the forward-reflection field. An especially interesting variety of camera is composed of two coaxial cones, one pointed toward the x-ray source and one in the opposite direction, as shown in Fig. 23. This records the entire field of Laue reflections.

In setting 5 the curves of constant ρ''''' and constant ϕ''''' are more complicated (Fig. 22*ii*).

Rösch[5] studied the geometry of the constant Υ and constant δ curves of the reflection sphere as projected on a cylinder when the incident x-ray beam coincides with the axis of the pole sphere and when this axis is normal to the incident beam. The charts were published in the first edition of the International Tabellen[11] without any correlation with Laue photographs. One of these charts, however, corresponds to the cylindrical Laue chart for setting 1, and the other to that of setting 2. Hermann and Ruhemann[10] have published curves of constant 2θ at intervals of 5° as projected onto

---

Fig. 21. Calcomp plot of the spot locations on a cylindrical Laue photograph of stearic acid showing the curves constant ϕ''' of setting 3.

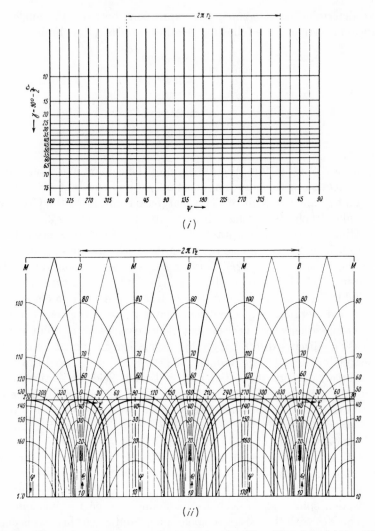

Fig. 22. Charts for constant ρ and constant ϕ for (i) setting 4 and (ii) setting 5. [From Schiebold and Schneider[11], p. 639, Fig. 22b, and 640, Fig. 23.]

cylindrical films; these correspond to the constant-ρ' curves of the cylindrical chart for setting 1.

The projection of the polychromatic component

In the foregoing sections, we have seen the relations existing between reference and reflection spheres, and between reflection sphere and film space. From those relations one can determine the spherical coordinates ρ, ϕ of the pole that corresponds to the Laue spot recorded in the film and from these we can attempt to interpret the Laue photograph.

The reference sphere gives the angles between poles; this set of poles is characteristic not only of the combination of external faces present in a given specimen but, since one Laue photograph provides the goniometry for all the crystal planes of a given substance, it also provides goniometric data for all crystals of the same substance. The reference sphere has a surface that cannot be flattened out on a sheet of paper, but it is possible to make various projections of it on a plane, as noted in Chapters 2, 3, and 4. The interpretation of a Laue diagram can hardly be made without a thorough knowledge of these methods of crystallographic projections. In order to interpret the Laue photograph, then, it is necessary to plot a projection of the reference sphere that corresponds to the reflection sphere of the actual Laue photograph, a problem that will be treated in the following sections.

The stereographic projection of the Laue photograph

The projection most suited to some aspects of the interpretation of the Laue photograph is the stereographic projection. In this case, the plane of the stereographic projection is the plane perpendicular to the x-ray beam passing through the center of the reference sphere and the viewpoint of the projection is the point of emergence of the primary x-ray beam from the reflecting sphere. The reference system of the angles ρ, ϕ of the pole that corresponds to a given Laue spot depends on the coordinate setting used. Therefore, in order to plot the stereographic projection of the reference sphere in connection with the Laue method, one must consider independently the two main coordinate settings that have been discussed in the foregoing sections.

Setting 1: polar stereographic net. In setting 1, the origin of ρ' is the point of incidence O' of the primary x-ray beam into the reflection sphere, and the origin of ϕ' is the equatorial great circle perpendicular to the axis of the camera, Fig. 24. From Chapter 2, the stereographic distance s from the center C of the primitive circle to the stereographic pole P_s is given by

$$s = R \tan \tfrac{1}{2}\rho' \tag{61}$$

where R is the radius of the primitive circle.

Since ϕ_s is measured on the primitive circle from the intersection of a given meridian circle, it is given by the angle between the reference meridian containing P. Therefore,

$$\phi_s = \phi'. \tag{62}$$

In the stereographic projection of this coordinate setting the curves of constant ϕ' are radii of the projection and the curves of constant ρ' are

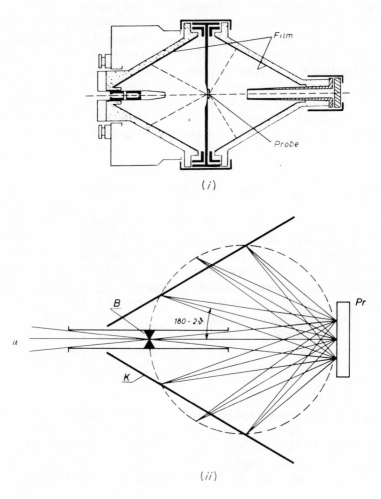

(*i*)

(*ii*)

Fig. 23. Conical camera for setting 4. (*i*) Section containing the x-ray beam. [From Herglotz[16], p. 620.] (*ii*) Focusing with the aid of a pinhole (second pinhole, D_2 of Figs. 1, 5, and 6, Chapter 13) whose location can be adjusted. [From Herglotz[20], p. 47.] (*iii*) Chart for interpreting the photograph of the back-reflection field in terms of coordinates θ ($\approx 90° - \rho'$) and δ ($\approx \theta$). [From Herglotz[20], p. 50, Fig. 11.]

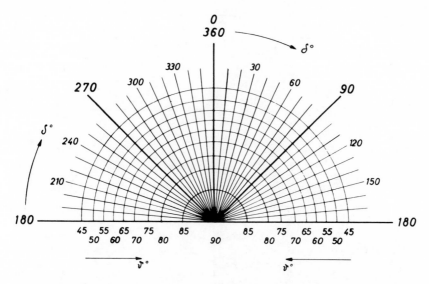

Fig. 23 (Continued)

small circles concentric on the center of the projection having radius s given by (61).

A chart constructed in such a way as to give curves of constant ϕ' and constant ρ' for definite (ordinarily every 2°) steps is called the *polar stereographic net* (Fig. 25).

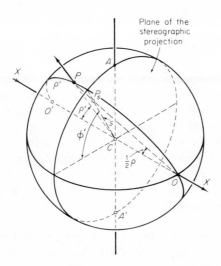

Fig. 24. Convention used for the stereographic projection of a Laue photograph.

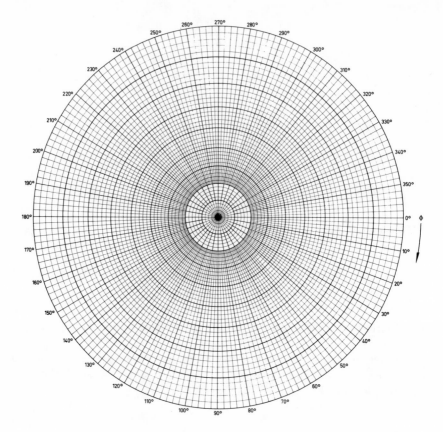

Fig. 25. Polar stereographic net; $R = 2$ inches.

Plotting the stereographic projection of a Laue photograph in setting 1.
Setting 1 has the advantage that the center of the projection is the origin
of ρ and therefore conforms with the normal crystallographic procedure.
In order to plot the stereographic projection in this case, the Laue spots
are numbered and the corresponding ρ', ϕ' coordinates are read as previously described. The plotting of the stereographic projection can be done
on the polar stereographic net of Fig. 25 in a direct way. The ρ' coordinate is read from the center of the projection and the ϕ' coordinate is read
along one radius containing the pole, ϕ' being measured in anticlockwise
sense.

Let us examine an actual example. A cylindrical Laue photograph of
triglycine sulfate mounted along [001] is given in Fig. 26. The angles
ρ', ϕ' of the different Laue spots have been measured utilizing the ρ', ϕ'

chart of Figure 8. With these values and the use of the polar stereographic chart the stereographic projection of Fig. 27 can be obtained. This normal procedure is useful when a few Laue spots are measured or when high-speed computers and plotters are not at hand.

The projection reproduced in Fig. 26 (as well as any others in this book) has been produced following the more practical scheme. The x, y coordinates of the Laue spots have been measured on the film with millimeter paper. These values are used as input to the computer program[23] STLPLT. This evaluates the ρ', ϕ' coordinates and the stereographic coordinates X_s, Y_s, which are, in turn, used to plot the projection in a CalComp plotter.

Blind areas. The use of a cylindrical film instead of a spherical one imposes restrictions on the x-ray reflection that can be received by the film. As a result, there are some empty areas on the pole sphere for a given orientation. The areas not recorded are called *blind areas*, and can readily be predicted by plotting in stereographic projection the geometrical limits of the film and the size and position on the film of the direct x-ray beam stop. The blind areas corresponding to the cylindrical-film case are shown in Fig. 28*i*. The outline corresponds with a film cylinder 80 mm long. The region covered by the beam stop appears in the projection as a limiting outer circle. The region not covered by the film, namely, the back slit and the top and bottom of the cylinder, is projected in the center of the projection. It can be seen that the territory covered by the blind areas in the case of a cylindrical camera is rather small and most of the projection area is available.

Because of the large recorded area, the cylindrical Laue method provides a means of studying the goniometry of a crystal with only one photograph. Once the poles have been plotted, the most important zones can be drawn easily with the aid of a Wulff net[3] (discussed in the next section) and the principal poles identified. The resulting figure will show the orientation of the crystal with respect to the x-ray beam. In addition, information on the crystal symmetry, discussed in Chapter 8, can be obtained.

A simulated flat transmission Laue photograph of triglycerine sulfate, having the same orientation as in the cylindrical Laue photograph of Fig. 26, is shown in Fig. 29. In the case of transmission Laue photographs, the ρ', ϕ' coordinates of the Laue spots can be read directly on the film with the help of the chart of Fig. 10. With those values, the stereographic projection can be plotted by using the polar stereographic chart. Figure 30 is the resulting stereographic projection.

Transmission flat Laue photographs can only register reflections from planes ranging from about $\rho = 60°$ to $\rho = 90°$, depending on the size of the film and the crystal-to-plate distance. The blind areas in this case are tre-

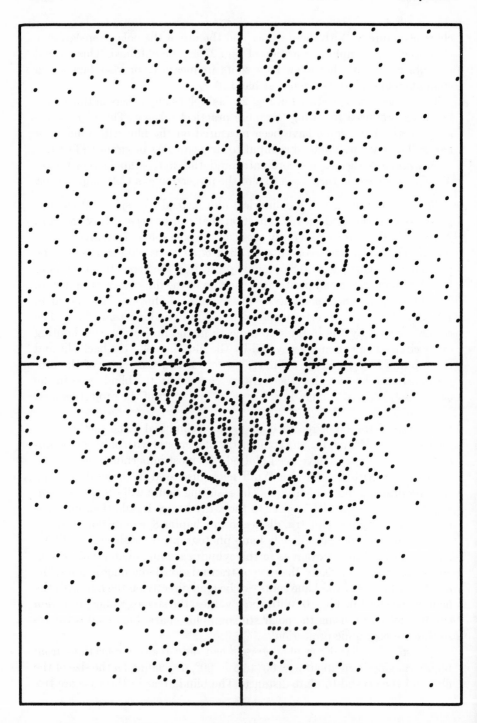

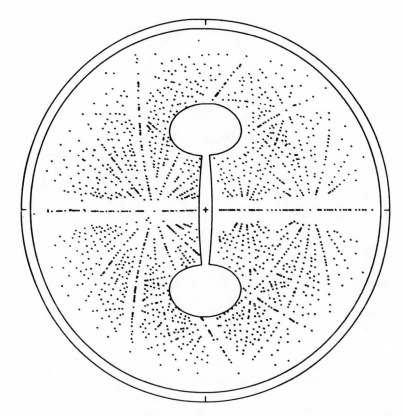

Fig. 27. Stereographic projection of the cylindrical Laue photograph given in Fig. 26. The blind areas are shown in the projection.

mendous, so that only a small peripheral region of the stereographic projection is covered by poles. The interpretation of the projection can be made only for simple cases, for example, for crystals with high symmetry that show several symmetry-related important poles.

A simulated back-reflection Laue photograph taken with the same crystal and orientation as in the cases previously mentioned is shown in Fig. 31. Following the technique described earlier, and using the chart reproduced in Fig. 10, the spherical coordinates ρ', ϕ' of the Laue spots can be determined and the stereographic projection shown in Fig. 32 can be plotted. In this case, only reflections of planes with poles in the range $\rho = 0°$ and

Fig. 26. CalComp plot of the spots on a cylindrical Laue photograph of triglycine sulfate mounted along [001]. Filtered copper x radiation has been used.

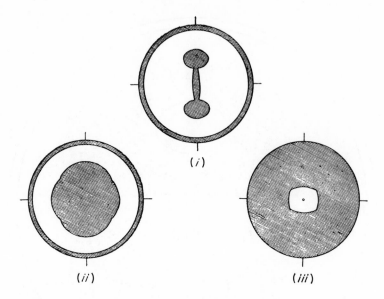

Fig. 28. Blind areas in the stereographic projections of Laue photographs. (*i*) Cylindrical; (*ii*) Transmission; (*iii*) Back-reflection. [After Canut and Amoros[18], p. 23.]

$\rho = 30°$ can be observed; the blind areas are even more evident in the corresponding stereographic projection than those of the transmission case. The projection in this particular case is difficult to interpret unless a thorough knowledge of the symmetry of the analyzed crystal is available. Where an unknown crystal is being studied, it is better to use the cylindrical-film method for obtaining Laue photographs. It is true, however, that when the crystal strongly absorbs the x rays (due to its size or high absorption coefficient), only reflections in the back-reflection region can be obtained, so this may limit the range recorded. The combination of the cylindrical Laue photograph and small crystals is recommended wherever possible, in order to avoid such disturbing effects.

Setting 2. The stereographic projection of the ρ'' and ϕ'' curves of setting 2 is more elaborate. The origin of ρ'' is the pole of the equatorial great circle which contains the direction of the x-ray beam and the axis of the camera (Fig. 11). The origin of ϕ'' is the primitive circle of the projection, that is, the principal circle perpendicular to the x-ray beam. The other great circles of constant ϕ'' are projected as circles whose radii and centers depend on the value of ϕ''.

Wulff net. Let us take $A'B'$ in Fig. 33 as the diameter of a great circle of constant ϕ''. We have

$$AO = R \tan \tfrac{1}{2}(90° - \phi''),\qquad(63)$$

$$OB = R \tan \tfrac{1}{2}(90° + \phi'').\qquad(64)$$

AB is the diameter of the projected circle of the great circle $A'B'$. The

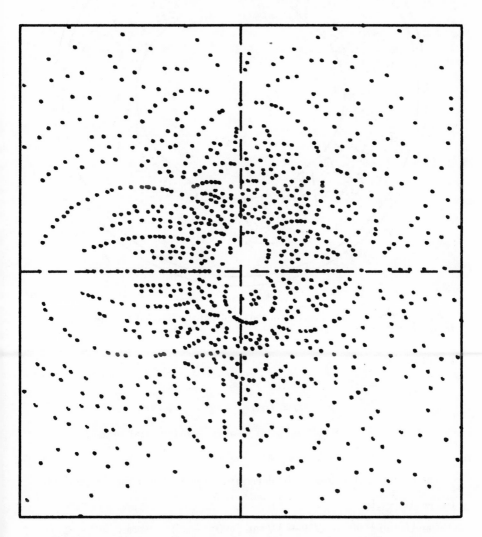

Fig. 29. Simulated transmission Laue photograph of triglycine sulfate. Crystal-to-film distance, 3 cm.

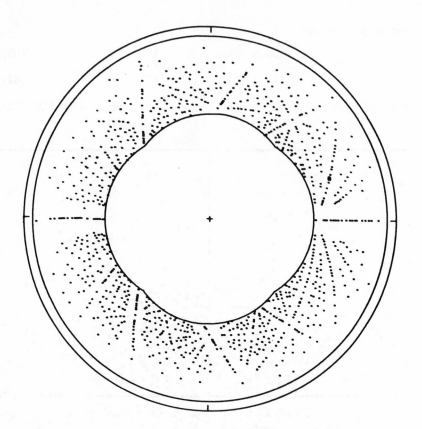

Fig. 30. Stereographic projection of the simulated transmission Laue photograph of triglycine sulfate crystal in Fig. 29.

radius of AB is

$$r = AC = AO + AC \tag{65}$$

$$= \tfrac{1}{2}(AO + OB)$$

$$= \tfrac{1}{2}R \tan \tfrac{1}{2}(90° - \phi'') + R \tan \tfrac{1}{2}(90° + \phi'')$$

$$= R \operatorname{cosec}(90° - \phi''). \tag{66}$$

The center of the circle is displaced from the center of the projection by

$$OC = AC - AO \tag{67}$$

$$= \tfrac{1}{2}R \tan \tfrac{1}{2}(90° + \phi'') + \tfrac{1}{2}R \tan \tfrac{1}{2}(90° - \phi'') - R \tan (90° - \phi'')$$

$$= \tfrac{1}{2}R \tan \tfrac{1}{2}(90° + \phi'') - \tfrac{1}{2}R \tan \tfrac{1}{2}(90° - \phi''). \tag{68}$$

In a simplified form,

$$OC = R \cot(90° - \phi'').\tag{69}$$

Small circles also project as circles in the stereographic projection. Of these, the important ones for the coordinate system are those which correspond to constant ρ'' and are also perpendicular to the plane of projec-

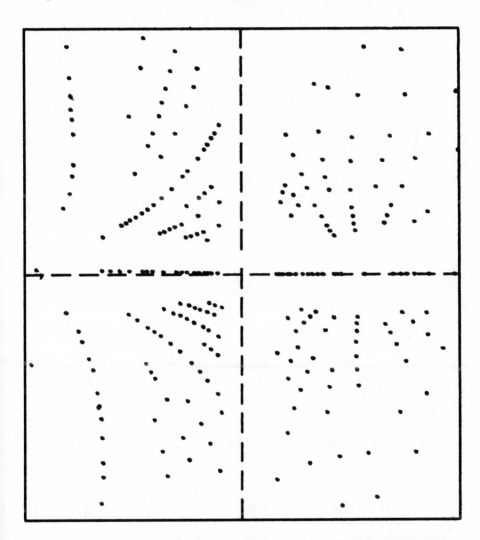

Fig. 31. Simulated back-reflection Laue photograph of triglycine sulfate. Crystal-to-film distance, 3 cm.

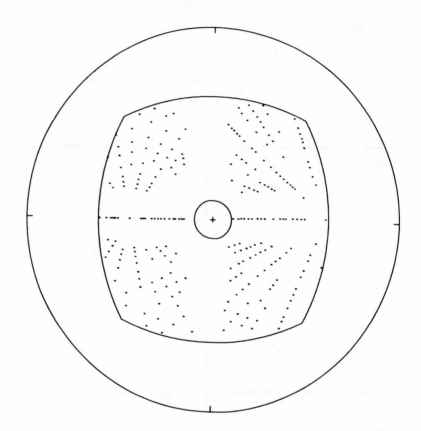

Fig. 32. Stereographic projection of the simulated back-reflection Laue photograph in Fig. 31.

tion. In Fig. 34, DD' is such a small circle. The projection of D is N and the projection of D' is M. The positions of N and M are the extremes of a diameter of the projected circle, and the center of the circle is C'. From the figure we can deduce

$$OC' = ON + \tfrac{1}{2} NM = \frac{R}{2} \left[\tan \tfrac{1}{2}(90° - \rho'') + \tan \tfrac{1}{2}(90° + \rho'') \right] \quad (70)$$

and

$$NC' = OC' - ON = \frac{R}{2} \left[\tan \tfrac{1}{2}(90° + \rho'') - \tan \tfrac{1}{2}(90° - \rho'') \right]. \quad (71)$$

In a simplified form

$$OC' = R \operatorname{cosec}(90° - \rho'') \quad (72)$$

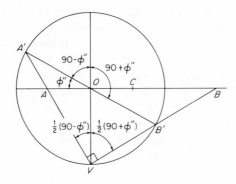

Fig. 33. Geometrical construction of great circles of the Wulff net.

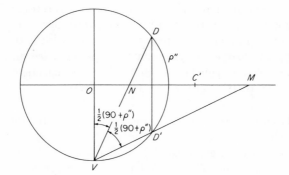

Fig. 34. Geometrical construction of small circles of the Wulff net.

and

$$r = R \cot(90° - \rho'').$$ (73)

Equations (67) and (69) can be used to construct a set of great circles evenly inclined around a given diameter of the primitive circle, while equations (72) an (73) are used to construct a set of small circles perpendicular to that diameter and evenly spaced. The whole figure so constructed contains two sets of circles, or arcs of circles, each having collinear centers, the lines of the centers being normal to each other, through the center O of the projection circle of radius R. A chart that contains these curves is known as the *Wulff net*[3], which was discussed in Chapter 2.

Plotting the stereographic projection of a Laue photograph in setting 2. A procedure similar to that used for setting 1 can be used for setting 2. In this case the center of the projection is again the direction of the x-ray beam but the origin of ρ'' is the north pole of the projection. The angular coordinates ρ'', ϕ'' of the Laue spot must be determined in the photograph with the chart of Henry, Lipson and Wooster[15]; the Wulff net may then be used as an aid in plotting the stereographic projection. If the flat-film technique is used, the ρ'', ϕ'' coordinates can be read directly from the film by using the Greninger[12] and Leonhardt[4] charts for transmission and reflection techniques, respectively. Again the Wulff net is used for the actual plotting of the projection.

If such procedures are used, one obtains a unique solution for any Laue photograph. However, if the wrong combination of charts and plotting is used, the answer may not be unique.

Gnomonic projection of a Laue photograph

The gnomonic projection can also be used for interpreting a Laue photograph. The plane of projection is a plane perpendicular to the x-ray beam and tangent to the reference sphere at the entrance point of the x-ray beam. The viewing point is, of course, the center of the reference sphere. The projection of the curves of constant ρ and constant ϕ depends on the setting, as does the actual plotting of the gnomonic projection.

Setting 1: the polar gnomonic net. In this setting the origin of ρ' is the point where the x-ray beam enters the reference sphere. In this case the very simple relation given by (1) in Chapter 3 should be used.

The curves of constant ρ' on the reference sphere are small circles; their gnomonic projections are a set of circles concentric with the center of the projection. The radii of such projected circles are given by (1), Chapter 3. The ϕ' angles are projected at their actual value, because the curves of constant ϕ' are great circles passing through the point of tangency of the projection plane with the pole sphere: the projections of such great circles are radial lines passing through the center of the gnomonic projection. When ρ' exceeds about 75°, however, it becomes impracticable to plot a pole, and this is one of the most important disadvantages of the gnomonic projection. Poles on the primitive circle of the stereographic projection, for which $\rho' = 90°$, are at infinity in the gnomonic projection, so they cannot be plotted, but they are ordinarily represented by arrows drawn around the rim of the plot, and pointing in the direction of the face normal. Besides this, poles of the hemisphere opposite to the point of tangency of the projection plane cannot be projected at all. This is not important in the interpretation of a Laue photograph, because use is made of only half of the pole sphere anyway, as we have already seen.

In a way similar to that used to interpret the stereographic projection, a polar gnomonic chart (Chapter 3, Fig. 17) can be used to plot the poles in a direct way once the ρ', ϕ' coordinates of the poles of the Laue spots have been determined by using the appropriate charts. Some inconveniences arise, however; first, because of the restrictions just explained about the practicality of the projection of poles with ρ' greater than 75°, the use of the gnomonic projection for cylindrical and transmission Laue photographs is not generally recommended. For the sake of completeness, Fig. 35 reproduces the gnomonic projection corresponding to the cylindrical Laue photograph of triglycine sulfate shown in Fig. 26. The boundary circle corresponds to $\rho' = 71.0°$. In a similar way, Figs. 36 and 37 reproduce the gnomonic projections corresponding to the stereographic projections of Figs. 30 and 32. Despite its inherent limitations, the gnomonic projection

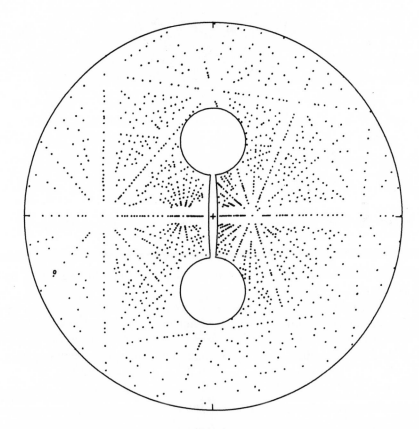

Fig. 35. Gnomonic projection of a cylindrical Laue photograph of triglycine sulfate made with the x-ray beam along [001], as in Fig. 31. The boundary circle is $\rho' = 71°$.

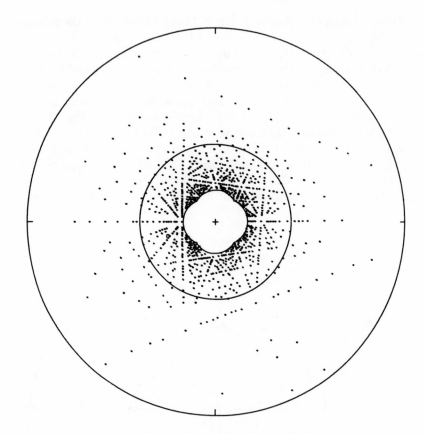

Fig. 36. Gnomonic projection of the simulated transmission Laue photograph of triglycine sulfate shown in Fig. 28. The boundary circle is $\rho' = 84°$.

is so easy to construct from a flat Laue photograph that plotting it may be worthwhile.

Let us consider first the transmission case. We assume NN' in Fig. 38i to be the plane of the gnomonic projection perpendicular to the x-ray beam and tangent to the reflection sphere whose center is C. A plane E will give a reflection L according to the Bragg equation. The corresponding pole of the plane E in the gnomonic projection is P_g. These relations follow from the figure:

$$OL = R \tan 2\theta, \tag{74}$$

$$OP_g = R \cot \theta \tag{75}$$

where R is the radius of the projection sphere. The same relations can be obtained for the flat Laue back-reflection case (Fig. 38*ii*). Accordingly, the interpretation of a conventional Laue photograph is straightforward. The gnomonic pole of the plane that gives rise to a given Laue spot is on the straight line that connects the Laue spot and the center of the Laue photograph. The pole is at a distance given by (75). The transmission technique differs from the back-reflection technique only in that, with the former method the gnomonic pole is not in the same part of the center of the Laue photograph as the Laue spot, whereas with the latter it is.

Setting 2: the Hilton net. In this setting the origin of ρ'' is taken at the north pole of the reference sphere, Fig. 11. The plane of the projection is the same as in setting 1. The curves of constant ρ'' in the reference sphere

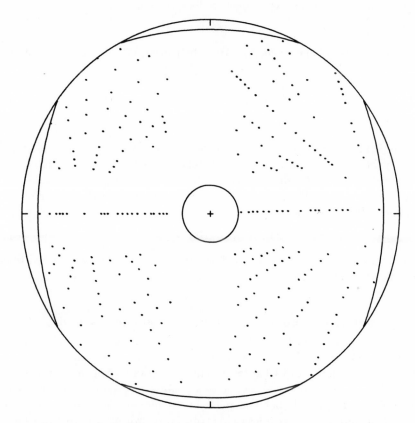

Fig. 37. Gnomonic projection of the simulated back-reflection Laue photograph shown in Fig. 30. The boundary circle is $\rho' = 34°$.

are given by a set of small circles perpendicular to a line parallel to the plane of projection. Their projection is a set of hyperbolas referred to their own axes whose equations are

$$\frac{x^2}{a^2} - \frac{y^2}{b^2} = 1.$$ (76)

By simple geometry, it can be deduced that in this case

$$a = R \cot \rho'' \quad \text{and} \quad b = R.$$ (77)

By substituting in (76) the values of (77), one obtains

$$\frac{x^2}{R^2 \cot^2 \rho''} - \frac{y^2}{R^2} = 1.$$ (78)

The focal points of the hyperbolas are located in a line, at distances

$$c = (a^2 + b^2)^{1/2}$$ (79)

from the center of the projection. By substituting (77) in (79) one obtains

$$c = (R^2 \cot^2 \rho'' + R^2)^{1/2}$$

$$= R(1 + \cot^2 \rho'')^{1/2}$$

$$= R \operatorname{cosec} \rho''.$$ (80)

The set of curves of constant ϕ'' on the polar sphere is a set of great circles evenly inclined around a common diameter to which the small circles of constant ρ'' are perpendicular. This set of great circles projects as a set of parallel lines. The distance of the corresponding line to the center of the gnomonic projection is

$$OD = R \tan(90° - \phi'').$$ (81)

With these expressions a net called the *Hilton net* can be constructed that aids plotting of the gnomonic projection as indicated in Chapter 3.

Stereognomonic projection of a Laue photograph

Setting 1: the polar stereognomonic net. The stereognomonic projection was explained in Chapter 4. The projection of the system of ρ', ϕ' curves on the pole sphere gives rise to a *polar stereognomonic net* similar to the polar stereographic net and polar gnomonic net. The small circles of $\rho' = $ constant are projected as circles in both projections. In the stereognomonic case they are, therefore, also projected as circles. The only dif-

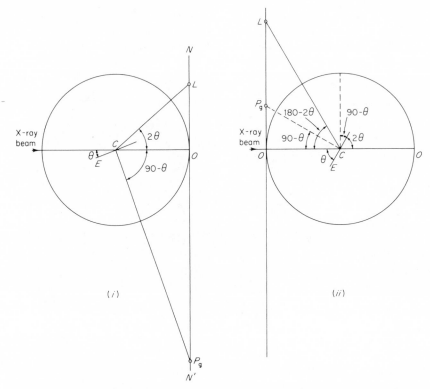

Fig. 38. Relation between Laue spot and gnomonic pole in conventional Laue photographs. (*i*) Transmission case. (*ii*) Back-reflection case.

ference is that for $\rho' <$ concordance angle, the radius of each circle is given by the gnomonic relation

$$r = R_g \tan \rho' \tag{82}$$

where R_g is the gnomonic radius (see Chapter 4). For $\rho >$ concordance angle, the radii of the circles are given by the stereographic relation

$$r' = R_s \tan \tfrac{1}{2}\rho' \tag{83}$$

where R_s has the same value as in the normal stereographic projection, namely 10 cm.

In setting 1 the great circles of constant ϕ' are projected in both projections along identical straight lines. Therefore the lines of constant ϕ' in the stereognomonic projection are radii of the projection circle (Fig. 9, Chapter 4). Figure 39 represents the stereognomonic projection of the cylindrical Laue photograph of triglycine sulfate of Fig. 26.

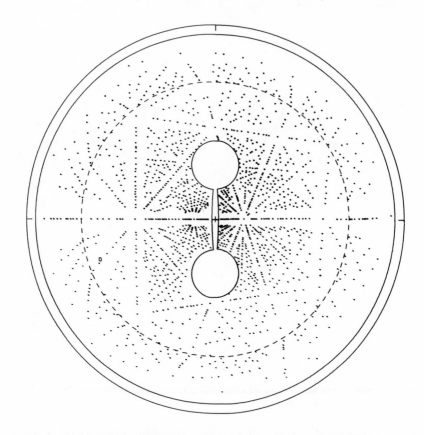

Fig. 39. Stereognomonic projection of the cylindrical Laue photograph of Fig. 25. The dashed concordance circle is at $\rho = 73°$.

Setting 2: the Parker net. Obviously the projection of the system of ρ'', ϕ'' curves for setting 2 in the stereognomonic projection is similar to the Wulff and Hilton nets. In fact, it is built up of both of them in the following way. The stereognomonic projection of the great circles of constant ϕ'' now have a mixed character. If ϕ'' is smaller than 90° minus the concordance angle, or greater than 90° plus the concordance angle, the curves of constant ϕ'' are exactly those of the stereographic Wulff net. For any other value the curves are circles for the part with $\rho'' > 90° - C$ and straight lines for the part with $\rho'' < 90° - C$ where C is the concordance angle.

In a similar way, the circles of constant ρ'' are projected as circles in

$\rho'' < 90° - C$. For all other cases, the lines of constant ρ'' are now part of circles or parabolas, depending on the value of ϕ'' (see Fig. 8, Chapter 4).

Significant literature

1. S. L. Penfield. The stereographic projection and its possibilities, from a graphical standpoint. *Am. J. Sci.* **11** (1901) 1–24, 115–144.
2. S. L. Penfield. Ueber die anwendung der stereographischen Projection. *Z. Kristallogr.* **35** (1901) 1–24.
3. Georg Wulff. Untersuchung im Gebiete der optischen Eigenschaften isomorpher Kristalle. *Z. Kristallogr.* **76** (1902) 1–28, especially 9–10, 14–18, Taf. II.
4. J. Leonhardt. Die Deutung der Lauediagramme deformierte Kristalle. *Z. Kristallogr.* **61** (1924) 100–112.
5. Siegfried Rösch. Ein graphischer Projektions-Transporteur. *Z. Kristallogr.* **64** (1926) 76–78.
6. J. Leonhardt. Anwendung der Zylinderprojektion bei der Lösung Kristallographischer Aufgaben. *Z. Kristallogr.* **76** (1930) 252–260.
7. F. Regler. Neue Methode zur Untersuchung von Faserstrukturen und zum Nachweis von inneren Spannungen an technischen Werkstücken. *Z. Phys.* **71** (1931) 371–388, especially 382–386.
8. F. Regler. Über eine neue Methode zur vollständigen röntgenographischen Feinstruktur-Untersuchung an technischen Werkstücken. *Z. Phys.* **74** (1932) 547–564.
9. E. Schiebold. Die Lauemethode. Vol. 1 of "Methoden der Kristallstrukturbestimmung mit Röntgenstrahlen" (Akademische Verlagsgesellschaft, Leipzig, 1932) 173 pages.
10. C. Hermann and M. Ruhemann. Die Kristallstruktur von Quecksilber. *Z. Kristallogr.* **83** (1932) 136–140.
11. E. Schiebold and E. Schneider. Graphical methods of evaluating x-ray diagrams. Internationale Tabellen zur Bestimmung von Kristallstrukturen. Vol. 2 (1935) 618–650.
12. Alden B. Greninger. A back-reflection Laue method for determining crystal orientation. *Z. Kristallogr.* **91** (1935) 424–432.
13. Antoni Laszkiewicz. Ueber die Zylinder-Laueaufnahmen. Archives de Minéralogie de la Société des Sciences et des Lettres de Varsovie [Poland] (1935) 105–112.
14. C. G. Dunn and W. W. Martin. The rapid determination of orientations of cubic crystals. *Metal Trans. AIME* **185** (1949) 417–427.
15. N. F. M. Henry, H. Lipson, and W. A. Wooster. The interpretation of x-ray diffraction photographs. (MacMillan, London, 1951) 258 pages, especially 108.
16. Herbert Herglotz. Das Doppelkegelverfahren und seine Eignung für Untersuchungen der Verformung und der Textur. *Z. Metallkunde* **46** (1955) 620–622.
17. B. D. Cullity. Elements of x-ray diffraction. (Addison-Wesley, Reading, Massachusetts, 1956) 514 pages, especially 220.
18. M. L. Canut and J. L. Amorós. Interpretacion racional de los Lauediagrams. *Bol. Real. Soc. Esp. Hist. Nat.* (**G**)**56** (1958) 15–24.

19. M. Canut. Tables for conversion of cylindrical Laue patterns to stereographic projections. [In: John S. Kasper and Kathleen Lonsdale (editors). "International tables for x-ray crystallography." Vol. II (1959)] 168–174.
20. H. K. Herglotz. Ein Röntgenverfahren zur Bestimmung der Fehlerhaftigkeit des Aufbaues von Einkristallen. *Acta Physica Austriaca* 15 (1962) 40–56.
21. A. Bernalte. On the curves in the Greninger and Leonhardt nets. *Acta Cryst.* 19 (1965) 916–918.
22. Leonid V. Azároff. Elements of x-ray crystallography. (McGraw-Hill, New York, 1968) 610 pages, especially 408.
23. Marisa Canut-Amoros. STLPLT—Calcomp plot of crystallographic projections of Laue photographs. *Computer Phys. Communications* 1 (1970) 293–305.

Chapter 10

The cross ratio and its application
in crystallography

Geometrical background

Projective geometry deals with the relations among a collection of geo-
metric elements that remain invariant when the collection of elements is
projected. A basic invariant relation is known as the *cross ratio* (sometimes
called the *double ratio, anharmonic ratio,* or *sine ratio.*) This has proven
useful in crystallography.

Invariance of cross ratio. A set of lines radiating from a point, such
as VM, VN, VP, and VQ in Fig. 1, is called a *pencil*. The common point
(V in Fig. 1) is called the *vertex* of the pencil and a line across the pencil,
like the line through A, B, C, D, is called a transversal. It is obvious that
the magnitudes of the corresponding segments of different transversals of a
pencil, like AB and $A'B'$ in Fig. 1, are not generally equal, nor are the
ratios of a pair of corresponding segments, like AB/BC and $A'B'/B'C'$.
But a remarkable feature is that a ratio of a pair of certain ratios is the same
for all transversals. For example, the ratio of the ratios

$$\frac{AC}{BC} \bigg/ \frac{AD}{BD} = r \tag{1}$$

is the same for all transversals. To prove this, the areas of certain triangles

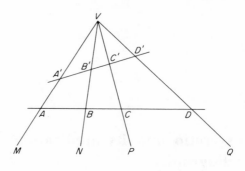

Fig. 1

in Fig. 1 are expressed in two ways, as suggested in Fig. 2:

$$\text{area } AVC = \tfrac{1}{2}h_V \cdot AC = \tfrac{1}{2}h_C \cdot VA = \qquad\qquad \tfrac{1}{2}h_A \cdot VC$$
$$= \tfrac{1}{2}(VC \sin \angle AVC) \cdot VA = \tfrac{1}{2}(VA \sin \angle AVC) \cdot VC$$
$$= \tfrac{1}{2}(VA \cdot VC) \sin \angle AVC.\dagger$$

In this way, four triangles with the common vertex V, and whose horizontal bases are the segments in (1), provide the four relations

$$\triangle AVC: \qquad \tfrac{1}{2}h_V \cdot AC = \tfrac{1}{2}VA \cdot VC \sin AVC; \qquad\qquad (2)$$

$$\triangle BVC: \qquad \tfrac{1}{2}h_V \cdot BC = \tfrac{1}{2}VB \cdot VC \sin BVC; \qquad\qquad (3)$$

$$\triangle AVD: \qquad \tfrac{1}{2}h_V \cdot AD = \tfrac{1}{2}VA \cdot VD \sin AVD; \qquad\qquad (4)$$

$$\triangle BVD: \qquad \tfrac{1}{2}h_V \cdot BD = \tfrac{1}{2}VB \cdot VD \sin BVD. \qquad\qquad (5)$$

When these are substituted in (1) it can be reduced as follows.

$$\dfrac{\dfrac{AC}{BC}}{\dfrac{AD}{BD}} = \dfrac{\dfrac{VC \cdot VA \sin AVC}{VC \cdot VB \sin BVC}}{\dfrac{VD \cdot VA \sin AVD}{VD \cdot VB \sin BVD}}, \qquad\qquad (6)$$

so that

$$\frac{AC}{BC} \Big/ \frac{AD}{BD} = \frac{\sin AVC}{\sin BVC} \Big/ \frac{\sin AVD}{\sin BVD}. \qquad\qquad (7)$$

$\dagger$ In vector algebra this is compactly expressed as $\tfrac{1}{2}\,\overrightarrow{VA} \times \overrightarrow{VC}$, but in the present application it is useful to evaluate the area of the triangle so that the sine of the angle between the two vectors is explicitly stated.

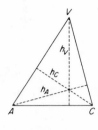

Fig. 2

An alternative form of this is

$$\frac{AC \cdot BD}{BC \cdot AD} = \frac{\sin AVC \cdot \sin BVD}{\sin BVC \cdot \sin AVD}. \tag{8}$$

This clearly shows that the cross ratio of the magnitudes of the line segments does not depend on the particular transversal, but is a function only of the sines of the angles at the vertex of the pencil. For example, it is the same for the transversals $ABCD$ and $A'B'C'D'$ in Fig. 1. The corresponding points in these two transversals are projections of one another from the vertex V. The sequence of points where the rays of the pencil cut the transversal is commonly called the range $ABCD$.

Invariance of the cross ratio with projection. Not only is the cross ratio independent of the transversal, but it is also true that, for a given range of points, the cross ratio of the sines of the angles is independent of the particular pencil, as illustrated in Fig. 3. This is because the cross ratio of sines on the right of (7) depends only on the signed magnitudes of the segments of the transversal, on the left of (7).

So far, cross ratios have been implicitly considered in connection with geometrical features in a plane. They can be extended to three dimensions

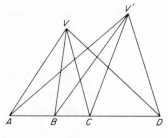

Fig. 3

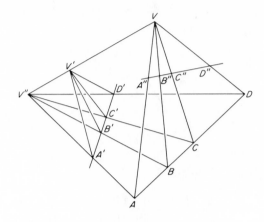

Fig. 4

as follows. In Fig. 4, consider a plane pencil VA, VB, VC, and VD and its transversal $ABCD$, all in a plane VAD. Draw a line from vertex V to any arbitrary point V'' not in plane VAD. Then pass four planes through the line VV'', one extending through A, one through B, one through C, and one through D. The points $V''VAD$ outline a tetrahedron. Pass an arbitrary plane $V'A'D'$ between vertex V'' and face VAD of this tetrahedron. Now consider the three pencils illustrated in Fig. 4, with vertices at V'', V', and V and the two transversals $ABCD$ and $A'B'C'D'$. The corresponding segments of these transversals have the same cross ratio, as illustrated in Fig. 1. Thus the cross ratios of the sines of corresponding angles V' and V are the same. But the pencil and transversal in plane VAD is a projection of the pencil and transversal in plane $V'A'D'$. It can be concluded that the cross ratios of a pencil and a transversal are invariant under projection.

Cross ratio of coaxial planes. The foregoing demonstration involved four collinear planes having the line $VV'V''$ of Fig. 4 in common, specifically the planes VAA', VBB', VCC', and VDD'. If another transversal is drawn in plane VAD, namely $A''B''C''D''$, it is obvious from Fig. 1 that this transversal also has the same cross ratio of segments as $A'B'C'D'$. So it can be concluded that all lines through the four planes have the same cross ratio, and that this cross ratio is dependent only on the angles between the planes. The cross ratio is called the cross ratio of four coaxial planes. This relation is important in crystallography because sets of coaxial planes occur in crystals, in reciprocal lattices, and in cylindrical and spherical coordinates to which the geometry of crystals is frequently referred.

Relations between cross ratios with different sequences. The cross ratio derived in (2)–(7) is not the only one that can be derived from Fig. 1. For example, an equally valid cross ratio can be based upon the areas of triangles AVB, BVC, AVD, and CVD, as follows:

$$\frac{\dfrac{AB}{BC}}{\dfrac{AD}{CD}} = \frac{\dfrac{VA \cdot VB \sin AVB}{VB \cdot VC \sin BVC}}{\dfrac{VA \cdot VD \sin AVD}{VC \cdot VD \sin CVD}} = \frac{\dfrac{\sin AVB}{\sin BVC}}{\dfrac{\sin AVD}{\sin CVD}}. \tag{9}$$

To survey all such possibilities by a scheme equivalent to (9) requires considering all the ways the limiting triangle AVD can be subdivided into smaller triangles whose bases are positive or negative segments of line AD.

An alternative method of surveying all possibilities is due to Möbius,[1] who thoroughly studied what he termed double-cut ratios (*Doppelschnittverhältnisse*) as early as 1827. Möbius considered a line within which were two segments AB and CD. The ends of each segment were regarded as cutting the other segment twice, either externally or internally. Möbius surveyed the various possible double-cut ratios by essentially adopting the following four rules.

1. Möbius defined his

$$Doppelschnittverhältnisse \equiv \frac{AC}{CB} : \frac{AD}{DB}. \tag{10a}$$

2. The points A, B, C, and D can be set down in any order. The number of orders is the numbers of permutation of four items, namely, $4 \cdot 3 \cdot 2 \cdot 1 = 24$ sequences.

3. The sign of a term is positive if the sequence of terms is alphabetic, but negative if it is the reverse of alphabetic.

4. An angle is positive if it increases in a counterclockwise sense, but negative if it increases in a clockwise sense.

If Möbius' double-cut ratio is written as the fraction

$$\frac{AC/CB}{AD/DB},$$

the numerator and denominator of the larger fraction can be represented for various sequences of letters by diagrams. For the sequence $A\ B\ C\ D$ the diagram is

$$A\ \overset{\frown}{B\ C}\ \underset{\smile}{D}$$

Table 1

Relations between values of cross ratios

Group number	Sequence symbol	Relative value	Anharmonic ratio when defined as $\dfrac{(1,\,3)}{(3,\,2)} \Big/ \dfrac{(1,\,4)}{(4,\,2)}$ of sequence (1, 2, 3, 4)
i	$(ABCD)$		$\dfrac{AC}{CB} \Big/ \dfrac{AD}{DB}$
	$(BADC)$		$\dfrac{BD}{DA} \Big/ \dfrac{BC}{CA}$
	$(CDAB)$	λ	$\dfrac{CA}{AD} \Big/ \dfrac{CB}{BD}$
	$(DCBA)$		$\dfrac{DB}{BC} \Big/ \dfrac{DA}{AC}$
ii	$(BACD)$		$\dfrac{BC}{CA} \Big/ \dfrac{BD}{DA}$
	$(ABDC)$		$\dfrac{AD}{DB} \Big/ \dfrac{AC}{CB}$
	$(CDBA)$	$\dfrac{1}{\lambda}$	$\dfrac{CB}{BD} \Big/ \dfrac{CA}{AD}$
	$(DCAB)$		$\dfrac{DA}{AC} \Big/ \dfrac{DB}{BC}$
iii	$(ACBD)$		$\dfrac{AB}{BC} \Big/ \dfrac{AD}{DC}$
	$(CADB)$		$\dfrac{CD}{DA} \Big/ \dfrac{CB}{BA}$
	$(BDAC)$	$1 - \lambda$	$\dfrac{BA}{AD} \Big/ \dfrac{BC}{CD}$
	$(DBCA)$		$\dfrac{DC}{CB} \Big/ \dfrac{DA}{AB}$

Table 1 (continued)

Group number	Sequence symbol	Relative value	Anharmonic ratio when defined as $\dfrac{(1,\ 3)}{(3,\ 2)} \Big/ \dfrac{(1,\ 4)}{(4,\ 2)}$ of sequence $(1, 2, 3, 4)$
iv	$(CABD)$		$\dfrac{CB}{BA} \Big/ \dfrac{CD}{DA}$
	$(ACDB)$	$\dfrac{1}{1-\lambda}$	$\dfrac{AD}{DC} \Big/ \dfrac{AB}{BC}$
	$(BDCA)$		$\dfrac{BC}{CD} \Big/ \dfrac{BA}{AD}$
	$(DBAC)$		$\dfrac{DA}{AB} \Big/ \dfrac{DC}{CB}$
v	$(BCAD)$		$\dfrac{BA}{AC} \Big/ \dfrac{BD}{DC}$
	$(CBDA)$	$1-\dfrac{1}{\lambda}=\dfrac{\lambda-1}{\lambda}$	$\dfrac{CD}{DB} \Big/ \dfrac{CA}{AB}$
	$(ADBC)$		$\dfrac{AB}{BD} \Big/ \dfrac{AC}{CD}$
	$(DACB)$		$\dfrac{DC}{CA} \Big/ \dfrac{DB}{BA}$
vi	$(CBAD)$		$\dfrac{CA}{AB} \Big/ \dfrac{CD}{DB}$
	$(BCDA)$	$\dfrac{\lambda}{\lambda-1}$	$\dfrac{BD}{DC} \Big/ \dfrac{BA}{AC}$
	$(ADCB)$		$\dfrac{AC}{CD} \Big/ \dfrac{AB}{BD}$
	$(DABC)$		$\dfrac{DB}{BA} \Big/ \dfrac{AC}{CA}$

in which the sign of a term in the double-cut ratio is suggested by the direction of the arrow. If points B and C of this sequence are interchanged, the numerator and denominator are represented by the diagram

$$A \ C \ B \ D.$$

It is evident that the interchange changes the sign of the ratio and changes the magnitudes as well.

Regardless of the lettering, let four points be ordered in a sequence 1, 2, 3, 4. Then the Möbius cross ratio can also be defined as

$$\text{cross ratio} \equiv \frac{(1, 3)}{(3, 2)} \bigg/ \frac{(1, 4)}{(4, 2)}. \tag{10b}$$

Here $(1, 3)$ represents the magnitude and sign of a line segment from point 1 to point 3, etc. If this definition is adopted, and the four letters A, B, C, D are assigned the 24 different orders noted above in 2, the cross ratios listed in the last column of Table 1 result.

It might be supposed that each of the 24 possible sequences has a different value of the ratio defined in (10a). Actually, there are only six distinct cross ratios for the 24 sequences, and these are all related. They are systematically tabulated in Table 1.

For the sequence A, B, C, D, the cross ratio was designated (A, B, C, D) by Möbius in 1827. This is now commonly abbreviated $(A \ B \ C \ D)$. The 24 possible sequence symbols are given in the second column of Table 1; they are grouped in six sets of four sequences such that the members of a set have the same value of the cross ratio.

Although Möbius defined his double-cut ratio in (10a) so that the denominator of each fraction composing the ratio is antialphabetic, and therefore negative, a simpler equivalent ratio has both denominators positive. In some of this chapter the cross ratio is accordingly taken as

$$\text{cross ratio} \equiv \frac{AC}{BC} \bigg/ \frac{AD}{BD}. \tag{10c}$$

The relative values of the cross ratios for different sequences can be deduced from the following theorems.

Theorem 1. If any two letters are interchanged, and at the same time the other two letters are interchanged, the cross ratio remains the same. (This accounts for the groupings into sets of four in Table 1.)

Proof: The possible interchanges are (a) $\widehat{ABCD}$; (b) $\widehat{ABCD}$; (c) $\widehat{ABCD}$. (This is the sequence of interchanges within each set of four in Table 1. Each set is centrosymmetrical, and each diagonal is composed of one letter.)

(a)
$$\frac{AC}{BC}\bigg/\frac{AD}{BD}\xrightarrow{\widehat{ABCD}}\frac{BD}{AD}\bigg/\frac{BC}{AC}=\frac{BD}{AD}\cdot\frac{AC}{BC}=\frac{AC}{BC}\bigg/\frac{AD}{BD}.$$

(b)
$$\frac{AC}{BC}\bigg/\frac{AD}{BD}\xrightarrow{\widehat{ABCD}}\frac{CA}{DA}\bigg/\frac{CB}{DB}=\frac{CA}{DA}\cdot\frac{DB}{CB}=\frac{AC}{BC}\cdot\frac{BD}{AD}=\frac{AC}{BC}\bigg/\frac{AD}{BD}.$$

(c)
$$\frac{AC}{BC}\bigg/\frac{AD}{BD}\xrightarrow{\widehat{ABCD}}\frac{DB}{CB}\bigg/\frac{DA}{CA}=\frac{DB}{CB}\cdot\frac{CA}{DA}=\frac{AC}{BC}\cdot\frac{BD}{AD}=\frac{AC}{BC}\bigg/\frac{AD}{BD}.$$

Theorem 2. If the first two letters are interchanged, the value of the cross ratio is inverted. (This accounts for the relation between groupings *i* and *ii*, between *iii* and *iv*, and between *v* and *vi* of Table 1.)

Proof:

$$\frac{AC}{BC}\bigg/\frac{AD}{BD}=\frac{AC}{BC}\cdot\frac{BD}{AD}\xrightarrow{\widehat{ABCD}}\frac{BC}{AC}\cdot\frac{AD}{BD}=1\bigg/\left(\frac{AC}{BC}\cdot\frac{BD}{AD}\right).$$

Theorem 3. If the middle two letters are interchanged, the value of the cross ratio r becomes $1-r$. (This accounts for the relation between groupings *i* and *iii*, between *ii* and *v*, and between *iv* and *vi* of Table 1.)

Proof:
$$AB+BC+CA=0.$$

Multiply by CD:
$$AB\cdot CD+BC\cdot CD+CA\cdot CD=0.$$

In second term, substitute $CD=AD-AC$,

in third term, substitute $CD=BD-BC$:
$$AB\cdot CD+BC(AD-AC)+CA(BD-BC)=0.$$

Expand:
$$AB\cdot CD+BC\cdot AD-\underline{BC\cdot AC}+CA\cdot BD-\underline{CA\cdot BC}=0.$$

Reduce:
$$AB\cdot CD+BC\cdot AD+CA\cdot BD=0.$$

[This last result is known as Euler's theorem.]†

Divide by $AD \cdot BC$:
$$\frac{AB \cdot CD}{AD \cdot BC} + \frac{BC \cdot AD}{AD \cdot BC} + \frac{CA \cdot BD}{AD \cdot BC} = 0.$$

$$-\frac{AB}{CB}\bigg/\frac{AD}{CD} + 1 \quad -\frac{AC}{BC}\bigg/\frac{AD}{BD} = 0.$$

$$-(ACBD) + 1 - (ABCD) = 0.$$

Treatment of points at infinity. When one of the points, say D, of the range $ABCD$ lies at an infinite distance from the other points on the transversal, then the second part of the cross ratio (10c)

$$(ABCD) = \frac{AC}{BC}\bigg/\frac{AD}{BD} \tag{11}$$

is indeterminate. But the behavior of the cross ratio can be studied by first placing D just to the positive side of C and then causing D to move toward infinity. As this proceeds, the part of the cross ratio containing D tends toward 1, and therefore the value of the entire cross ratio tends toward the part of (11) not containing D, specifically AC/BC.

Special values of the cross ratio. When the cross ratio (10c) has the value 1, then

$$\frac{AC}{BC}\bigg/\frac{AD}{BD} = 1,$$

so that

$$\frac{AC}{BC} = \frac{AD}{BD}. \tag{12}$$

This can occur either when

$$C = D \quad \text{(i.e., when } C \text{ and } D \text{ coincide)},$$

† This relation is symmetrical in the product of each side and its opposite partial median in the trigonal diagram:

or when
$$A = B \qquad \text{(i.e., when } A \text{ and } B \text{ coincide).}$$

When the cross ration (10c) has the value -1, then

$$\frac{AC}{BC} = -\frac{AD}{BD}. \tag{13}$$

An obvious way to satisfy (13) is to place D at infinity. Then, according to the discussion in the foregoing section, the part of the ratio containing D becomes 1; this leaves

$$\frac{AC}{BC} = -1,$$

so

$$AC = -BC = CB.$$

This is satisfied when C is halfway between A and B, as illustrated in Fig. 5. Alternatively, C can be placed at infinity, which calls for

$$\frac{BD}{AD} = -1.$$

This is satisfied when D is halfway between A and B. Furthermore, since

$$\frac{AC}{BC} \Big/ \frac{AD}{BD} = \frac{AC \cdot BD}{BC \cdot AD} = \frac{AC}{AD} \Big/ \frac{BC}{BD},$$

it is also possible to satisfy (12) by placing B at infinity and A halfway between C and D or by placing A at infinity and B between C and D. Thus an obvious set of solutions of (13) has one of the four points at infinity and the other three spaced at equal intervals. This situation is commonly encountered in crystallography.

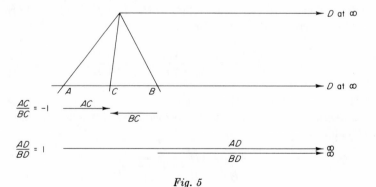

Fig. 5

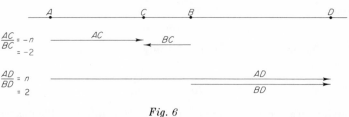

Fig. 6

A more general way to satisfy (13) is to fix the ratio AC/BC at some general value, say $-n$ (Fig. 6, which illustrates $n = 2$), and then find a location for D such that $AD/BD = +n$.

Applications in crystallography

One of the characteristic geometrical features of crystals is that a rational line occurs in many different rational planes, or, looked at from another viewpoint, that many rational planes have the same rational line in common. The planes of a set that have the same line in common are said to be *tautozonal planes*. The normals to the set of tautozonal planes all occur in the plane normal to the zone. This zone plane intersects the pole sphere in a great circle and along this great circle the poles appear as points. In the various projections, the poles appear as points along a line that is a straight line in the gnomonic projection, a great circle in the stereographic projection, and an ellipse or hyperbola in the Laue photograph. The sequence of poles along a line makes it possible to relate the angles between poles to such crystallographic features as the indices of the planes.

Form of cross ratio commonly adopted in crystallography. Crystallographers ordinarily use their own definition of "anharmonic ratio" which corresponds to the form derived in (9), specifically

(9): $$\frac{AB}{BC} \bigg/ \frac{AD}{CD} = \frac{\sin AVB}{\sin BVC} \bigg/ \frac{\sin AVD}{\sin CVD}.$$

The relation of this form to the Möbius form can be appreciated by setting down a diagram of the Möbius *Doppelschnittverhältnis* for the sequence *ABCD* as follows:

$$\frac{AC}{CB} \bigg/ \frac{AD}{DB} = A\,\overset{\frown}{B\,C}\,D,$$

and then relettering the sequence so that $C \to B'$ and $B \to C'$, thus:

$$\frac{AB'}{B'C'} \bigg/ \frac{AD}{DC'} = A\, B'\, C'\, D. \tag{14}$$

The left-hand sides of (9) and (14) are now similar except for the opposite signs of CD in (9) and DC' in (14).

Now, disregarding the primes on B and C, the definition of cross ratio for the left of (14) corresponds to

$$r_c \equiv \frac{(1,\,2)}{(2,\,3)} \bigg/ \frac{(1,\,4)}{(4,\,3)} = \frac{(1,\,2)}{(2,\,3)} \cdot \frac{(3,\,4)}{(4,\,1)}. \tag{15}$$

The right part of (15) has a simple cyclical sequence that is useful in certain crystallographic applications. While there are advantages in using the sequence of magnitudes in (15), there are also advantages in preserving the correct signs of the fractions, so (15) is accepted here as the crystallographer's cross ratio and this is identified by a subscript c.

Only when the signs of the segments in (15) are properly distinguished does the crystallographer's cross ratio bear a simple relation to any of the permutations, listed in Table 1, of the cross ratio based upon Möbius' definition (10a). This simple relation is

$$(ABCD)_c = (ACBD)_{\text{Möbius}}. \tag{16}$$

Nevertheless it should be clearly understood that crystallographers usually ignore the negative sign in the last fraction of (15). In this book the signs are carefully observed so that the results are more easily related to the standard mathematician's results.

Angles between poles in a zone. The cross ratio can be applied to easy graphical solutions involving angles between poles in a zone. In Fig. 7, four tautozonal poles in the sequence $ABCD$ are shown in stereographic projection. If O is the center of the sphere, then the cross ratio of the sines of the aperture angles, as set down in (8), is

$$(ABCD)_c = \frac{\sin AOC}{\sin BOC} \cdot \frac{\sin BOD}{\sin AOD}. \tag{17}$$

In the stereographic projection these angles can be measured by finding the stereographic projection Z of the pole of the zone circle containing A, B, C, and D, then drawing the lines ZA, ZB, ZC, and ZD (not shown in Fig. 7) until they meet the primitive circle; the arc of the primitive circle

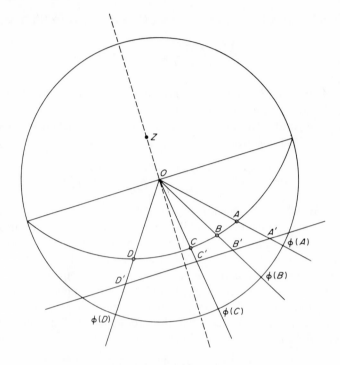

Fig. 7

subtended by the extension of ZA and ZC measures the angle between poles A and C, etc. In this way the angles in (17) can be evaluated.

But the invariance of the cross ratio under projection provides a shortcut. In Fig. 7, pass four planes normal to the primitive circle, each plane containing the center of the sphere, and one of the four poles A, B, C, or D. These planes intersect the sphere so that their stereographic projections are straight lines radiating from the center through A, B, C, and D. Along each line the coordinate ϕ is constant, so that the intersection of the line through A gives the ϕ coordinate of pole A, etc. Now, the projective invariance of the cross ratio of sines of four coaxial planes permits using, in (17), either the true angles between pairs of the poles A, B, C, and D, or the angles between the pairs of planes on which the poles lie. Thus, the differences in the ϕ coordinates between pairs of poles A, B, C, and D can be substituted in (17) for the angles themselves.

Cross ratio expressed as indices of planes. Cross ratios can be expressed in terms of the indices of four crystal planes in the same zone. In

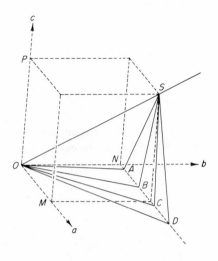

Fig. 8

Fig. 8, the directions of the crystallographic axes a, b, and c of an unspecialized crystal are shown. Some zone $[uvw]$ is indicated by the line OS which extends from the origin through a point whose coordinates on the crystallographic axes are $OM = ua$, $ON = vb$, and $OP = wc$. By definition, any planes in zone OS have this line in common. Such planes are OSA, OSB, OSC, and OSD. These intersect the ab plane in the pencil whose vertex is O and whose transversal contains the points $ABCD$.

The indices of planes OSA, OSB, OSC, and OSD are required. To determine these, translate NA, NB, NC, and ND along the direction $-b$, so that A, B, C, and D fall along the a axis. The new situation is illustrated in Fig. 9; in this drawing, the relative intercepts of the several planes on the a and b axes can be determined. Specifically

$$\frac{OA'}{OO'} = \frac{a/h_1}{-b/k_1}, \tag{18}$$

so that

$$OA' = -OO'\frac{a}{b} \cdot \frac{k_1}{h_1}; \tag{19}$$

similarly

$$OB' = -OO'\frac{a}{b}\cdot\frac{k_2}{h_2}, \tag{20}$$

$$OC' = -OO'\frac{a}{b}\cdot\frac{k_3}{h_3}, \tag{21}$$

$$OD' = -OO'\frac{a}{b}\cdot\frac{k_4}{h_4}. \tag{22}$$

The cross ratio (15) can be set down for sequence $A'B'C'D'$ of Fig. 9 and, after substitutions from (18)–(22), can be reduced as follows.

$$\frac{A'B'}{B'C'}\bigg/\frac{A'D'}{D'C'} = \frac{(OB'-OA')}{(OC'-OB')}\bigg/\frac{(OD'-OA')}{(OC'-OD')} \tag{23}$$

$$= \frac{-OO'\dfrac{a}{b}\left(\dfrac{k_2}{h_2}-\dfrac{k_1}{h_1}\right)}{-OO'\dfrac{a}{b}\left(\dfrac{k_3}{h_3}-\dfrac{k_2}{h_2}\right)}\bigg/\frac{-OO'\dfrac{a}{b}\left(\dfrac{k_4}{h_4}-\dfrac{k_1}{h_1}\right)}{-OO'\dfrac{a}{b}\left(\dfrac{k_3}{h_3}-\dfrac{k_4}{h_4}\right)}. \tag{24}$$

$$= \frac{\left(\dfrac{k_2}{h_2}-\dfrac{k_1}{h_1}\right)}{\left(\dfrac{k_3}{h_3}-\dfrac{k_2}{h_2}\right)}\bigg/\frac{\left(\dfrac{k_4}{h_4}-\dfrac{k_1}{h_1}\right)}{\left(\dfrac{k_3}{h_3}-\dfrac{k_4}{h_4}\right)} \tag{25}$$

$$= \frac{\dfrac{1}{h_1h_2}(h_1k_2-h_2k_1)}{\dfrac{1}{h_2h_3}(h_2k_3-h_3k_2)}\bigg/\frac{\dfrac{1}{h_1h_4}(h_1k_4-h_4k_1)}{\dfrac{1}{h_4h_3}(h_4k_3-h_3k_4)} \tag{26}$$

$$= \frac{\begin{vmatrix} h_1 & k_1 \\ h_2 & k_2 \\ h_2 & k_2 \\ h_3 & k_3 \end{vmatrix}}{\ }\bigg/\frac{\begin{vmatrix} h_1 & k_1 \\ h_4 & k_4 \\ h_4 & k_4 \\ h_3 & k_3 \end{vmatrix}}{\ }. \tag{27}$$

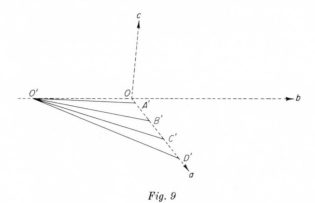

Fig. 9

Rearrangement of the left of (23) and the right of (27) yields the alternative form, analogous to (8):

$$\frac{A'B'}{B'C'} \cdot \frac{C'D'}{D'A'} = \frac{\begin{vmatrix} h_1 & k_1 \\ h_2 & k_2 \end{vmatrix}}{\begin{vmatrix} h_2 & k_2 \\ h_3 & k_3 \end{vmatrix}} \cdot \frac{\begin{vmatrix} h_3 & k_3 \\ h_4 & k_4 \end{vmatrix}}{\begin{vmatrix} h_4 & k_4 \\ h_1 & k_1 \end{vmatrix}}. \tag{28}$$

The theme in this derivation has been to find the indices of the four planes in the zone OS by the ratios of the intercepts of these planes on the a and b axes. This has been possible through the use of the cross ratio simply because the three indices of planes depend on the ratios of a/h, b/k, and c/l of a series of planes. In the derivation just given, only the a/h and b/k ratios of the four planes were involved, so that only the hk indices of the four planes were fixed.

But a result like (28) can be derived for each pair of indices hk, hl, and kl. For example, to derive the relation for the pair hl, let the planes in the zone OS indicated in Fig. 8 be extended to the other side of their common line OS, so that their traces on the axial plane OPM can be represented as drawn in Fig. 10. The slopes of these traces can be used to establish the ratios a/h and c/l of the four planes. To do this, use is made of the projection invariance of the cross ratio of a pencil and transversal. In particular, in Fig. 8, the sines of the aperture angles at S are the same as at O, since they have the common transversal $ABCD$. Furthermore, in Fig. 10, ranges $ABCD$ and $A''B''C''D''$ are the same, so that the cross ratio of the sines of their aperture angles from the vertex O are the same. Just as the ab plane of Fig. 8 was translated in the direction $-b$ so that N came to O and

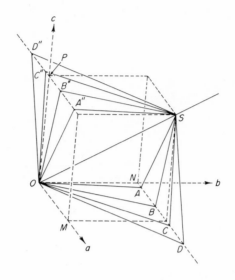

Fig. 10

O to O' in Fig. 9, so the ac plane of Fig. 10 is translated in the direction $-c$ so that P comes to O and O to O'' in Fig. 11. A procedure analogous to (23)–(28) then provides the relation between indices h and l of the four planes. A similar procedure utilizing the bc plane provides the relation between indices k and l.

If the four planes A, B, C, and D are designated 1, 2, 3, and 4, respec-

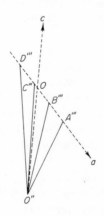

Fig. 11

tively, and when (8) as well as (28) and analogous results for other pairs of indices are taken into account, the following three relations are obtained:

$$\frac{\sin(1,2)}{\sin(2,3)} \cdot \frac{\sin(3,4)}{\sin(4,1)} = \frac{\begin{vmatrix} h_1 & k_1 \\ h_2 & k_2 \end{vmatrix}}{\begin{vmatrix} h_2 & k_2 \\ h_3 & k_3 \end{vmatrix}} \cdot \frac{\begin{vmatrix} h_3 & k_3 \\ h_4 & k_4 \end{vmatrix}}{\begin{vmatrix} h_4 & k_4 \\ h_1 & k_1 \end{vmatrix}}, \tag{29}$$

$$= \frac{\begin{vmatrix} h_1 & l_1 \\ h_2 & l_2 \end{vmatrix}}{\begin{vmatrix} h_2 & l_2 \\ h_3 & l_3 \end{vmatrix}} \cdot \frac{\begin{vmatrix} h_3 & l_3 \\ h_4 & l_4 \end{vmatrix}}{\begin{vmatrix} h_4 & l_4 \\ h_1 & l_1 \end{vmatrix}}, \tag{30}$$

$$= \frac{\begin{vmatrix} k_1 & l_1 \\ k_2 & l_2 \end{vmatrix}}{\begin{vmatrix} k_2 & l_2 \\ k_3 & l_3 \end{vmatrix}} \cdot \frac{\begin{vmatrix} k_3 & l_3 \\ k_4 & l_4 \end{vmatrix}}{\begin{vmatrix} k_4 & l_4 \\ l_1 & l_1 \end{vmatrix}}. \tag{31}$$

A simple mnemonic for these relations is to ignore one column of h or k or l in the general formula

$$\frac{\sin(1,2)}{\sin(2,3)} \cdot \frac{\sin(3,4)}{\sin(4,1)} = \frac{\begin{vmatrix} h_1 & k_1 & l_1 \\ h_2 & k_2 & l_2 \end{vmatrix}}{\begin{vmatrix} h_2 & k_2 & l_2 \\ h_3 & k_3 & l_3 \end{vmatrix}} \cdot \frac{\begin{vmatrix} h_3 & k_3 & l_3 \\ h_4 & k_4 & l_4 \end{vmatrix}}{\begin{vmatrix} h_4 & k_4 & l_4 \\ h_1 & k_1 & l_1 \end{vmatrix}}. \tag{32}$$

This result is useful in solving a problem involving angles between the four poles and the stereographic locations of the four poles when one of these items is unknown. The same problem can also be solved by a simple graphical method using the projection invariance of the cross product. To treat the problem graphically, a transversal is drawn across the pencil radiating from O in Fig. 7. The magnitudes of the segments can be easily measured by any convenient scale and the cross ratio computed. This can be converted to the sines of the true angles between poles, or the differences between ϕ coordinates of the poles.

Cross ratio expressed as distances from one point of the range.
The cross ratio of intervals between points of a range can be transformed
into a ratio of distances from a particular point of the range. Let the sequence of points on a transversal be $ABCD$. Then the cross ratio of $(ABCD)$
can be set down from (15) and then related to the sines in the pencil (with
due regard to the sign of DA) as

$$\frac{p}{q} = \frac{AB}{BC} \cdot \frac{CD}{DA} = \frac{\sin AB}{\sin BC} \cdot \frac{\sin CD}{\sin DA}, \tag{33}$$

in which $\sin AB$ is an abbreviation of $\sin AVB$, etc. The intervals can be
expressed as their distances from A, as follows.

$$\begin{aligned}
AB &= AB, \\
BC &= AC - AB, \\
CD &= AD - AC, \\
DA &= -AD.
\end{aligned} \tag{34}$$

With these substitutions, the right side of (33) can be reduced as follows.

$$\frac{\sin AB}{\sin BC} \cdot \frac{\sin CD}{\sin DA} = \frac{\sin AB \sin(AD - AC)}{\sin(AC - AB) \sin(-AD)}$$

$$= -\frac{\sin AB (\sin AD \cos AC - \cos AD \sin AC)}{(\sin AC \cos AB - \cos AC \sin AB) \sin DA}$$

$$= -\frac{(\sin AD \cos AC - \cos AD \sin AC)/\sin AD}{(\sin AC \cos AB - \cos AC \sin AB)/\sin AB}$$

$$= -\frac{\cos AC - \cot AD \sin AC}{\sin AC \cot AB - \cos AC} \quad \left[\times \frac{1/\sin AC}{1/\sin AC} \right]$$

$$= -\frac{\cot AC - \cot AD}{\cot AB - \cot AC}. \tag{35}$$

(If the correct sign of DA in (33) is ignored, the sign of the right of (35)
is positive.)

When this is equated to p/q of (33), the result can be reduced as follows:

$$\frac{p}{q} = -\frac{\cot AC - \cot AD}{\cot AB - \cot AC}, \tag{36}$$

$$p \cot AB - p \cot AC = -q \cot AC + q \cot AD,$$

$$p \cot AB - (p - q) \cot AC - q \cot AD = 0. \tag{37}$$

Table 2

Cross ratios and other relations expressed in terms of angular distances from specific poles in a zone

Pole from which angles in the zone are measured — Symbol	Position in sequence	Cross ratio	Cotangent formula (neglect of negative sign on right of cross ratio produces relation in parentheses)
A	1	$\dfrac{p}{q} = -\dfrac{\cot AC - \cot AD}{\cot AB - \cot AC}$	$p \cot AB - (p - q) \cot AC - q \cot AD = 0$ $(p \cot AB - (p + q) \cot AC + q \cot AD = 0)$
B	2	$\dfrac{p}{q} = -\dfrac{\cot BC - \cot BD}{\cot BD - \cot BA}$	$q \cot BC + (p - q) \cot BD - p \cot BA = 0$ $(-q \cot BC + (p + q) \cot BD - p \cot BA = 0)$
C	3	$\dfrac{p}{q} = -\dfrac{\cot CA - \cot CB}{\cot CD - \cot CA}$	$p \cot CD - (p - q) \cot CA - q \cot CB = 0$ $(p \cot CD - (p + q) \cot CA + q \cot CB = 0)$
D	4	$\dfrac{p}{q} = -\dfrac{\cot DA - \cot DB}{\cot DB - \cot DC}$	$q \cot DA + (p - q) \cot DB - p \cot DC = 0$ $(-q \cot DA + (p + q) \cot DB - p \cot DC = 0)$

Similar results may be derived in terms of distances from other points in the range. Several results are listed in Table 2. These relations are suited to the computation of angles between poles when the indices of four poles and three angles are known. The indices provide the cross ratio from (32), from which p/q can be calculated. The unknown angle can thus be computed from one of the relations in Table 2.

Schiebold's method of indexing. In Chapter 8 the indexing of the conventional Laue photograph of a cubic crystal with the aid of the gnomonic projection was discussed. The procedure is quite simple if the direction of the x-ray beam is such that the normal pattern of the gnomonic projection is obtained. Other specialized directions of the x-ray beam, such as along [100] and [010], also give rise to patterns in the gnomonic projection which can be easily recognized by taking account of basic principles.

If the incident x-ray beam is not parallel to a symmetry axis, or to a plane of the Friedel class, the gnomonic projection is asymmetric. It is still possible, however, to recognize the network defined by systems of important zones. A complication arises from the distortion introduced in the projection because, in general, the axial directions are not parallel to the plane of projection. In this case, the lengths that define the poles in the gnomonic projection are functions of the perspective angle. However, a relatively simple case is obtained when the x-ray beam is perpendicular to a plane (hkl). The gnomonic projection contains, at finite distances from the origin, the poles (100), (010), and (001), or their negatives, as shown in Fig. 12. The three axial poles define a triangle with the pole of (hkl) in its interior.

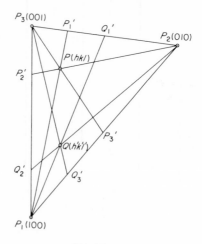

Fig. 12

From one axial pole, say P_1, Fig. 12, zone lines radiate. These radiating lines are the same as the set of lines that, in a normal projection, would be parallel. A line from P_2 or P_3 can likewise define a set of lines that would be parallel in a normal projection. Each such line is the gnomonic representation of a zone.

The poles $P_1(100)$, $P_2(010)$, and $P_3(001)$ are the vertices of a fundamental triangle of simple zones. Zone lines can be drawn from each of these poles through $P(hkl)$ and $Q(h'k'l')$. The intersections of these zones with the sides of the triangle at P_1', P_2', P_3' and Q_1', Q_2', and Q_3' are also possible poles. These points divide each side of the fundamental triangle into segments to which the cross ratio can be applied. For example, the fundamental triangle is shown in Fig. 13 in the form of a pencil with vertex P_2 and transversal $P_1Q_2'P_2'P_3$. The following treatment is due to Schiebold[13]. The cross ratio is

$$r = \frac{AC}{BC} \bigg/ \frac{AD}{DB} = \frac{\sin AVC}{\sin BVC} \bigg/ \frac{\sin AVD}{\sin DVB}. \tag{38}$$

The cross ratio in terms of indices is

$$
\begin{aligned}
r_1 &= \frac{k_1l_3 - l_1k_3}{k_2l_3 - l_2k_3} \bigg/ \frac{k_1l_4 - l_1k_4}{k_2l_4 - l_2k_4} \\[2mm]
&= \frac{l_1h_3 - h_1l_3}{l_2h_3 - h_2l_3} \bigg/ \frac{l_1h_4 - h_1l_4}{l_2h_4 - h_2l_4} \\[2mm]
&= \frac{h_1k_3 - k_1h_3}{h_2k_3 - k_2h_3} \bigg/ \frac{h_1k_4 - k_1h_4}{h_2k_4 - k_2h_4}.
\end{aligned}
\tag{39}
$$

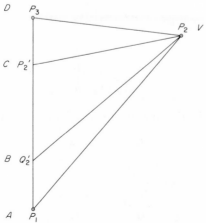

Fig. 13

In the case being considered, since the plane belongs to the zone $[100]$ and hence must have $k = 0$, the symbols of the planes are very simple, namely $P_1(100)$, $Q_2'(h_2 0 l_2)$, $P_2'(h_3 0 l_3)$, and $P_3(001)$. Accordingly, (39) is greatly simplified, leading to

$$r_1 = \frac{1}{1 - q_1} \tag{40}$$

where

$$q_1 = \frac{h_3}{l_3} \cdot \frac{l_2}{h_2}. \tag{41}$$

Because P_2', $-P(hkl)$, and (010) belong to the same zone, the ratio h_3/l_3 is identical with h/l of the plane (hkl) normal to the incident beam. Since hkl is known, (40) allows l_2/h_2 of (41) to be determined.

In a similar way, in the zone $P_2 Q_1' P_1' P_3$ the corresponding relation can be deduced as follows.

$$r_2 = \frac{1}{1 - q_2} \tag{42}$$

where

$$q_2 = \frac{k_3}{l_3} \cdot \frac{l_2}{k_2}. \tag{43}$$

Here $h_2 k_2 l_2$ refers to the symbol of Q_1, and $h_3 k_3 l_3$ corresponds to P_1'.

Also, for the zone $P_1 Q_3' P_3' P_2$ the corresponding results are

$$r_3 = \frac{1}{1 - q_3} \tag{44}$$

where

$$q_3 = \frac{h_3}{k_3} \cdot \frac{k_2}{h_2}. \tag{45}$$

From (39) and (37) the unknown symbol $(h_2 k_2 l_2)$ can be obtained:

$$\frac{l_2}{h_2} = \frac{l_0}{h_0} \cdot \frac{r_1 - 1}{r_1} ; \tag{46}$$

$$\frac{h_2}{k_2} = \frac{h_0}{k_0} \cdot \frac{r_2 - 1}{r_2} ; \tag{47}$$

$$\frac{l_2}{k_2} = \frac{l_0}{k_0} \cdot \frac{r_2 - 1}{r_3} .$$ (48)

The construction just explained allows one to determine the symbol of any Laue spot. In the case where two or more zones intersect in a point, the (hkl) symbol of the corresponding pole can be deduced in the usual manner.

Summary

A basic tool of projective geometry is known to mathematicians as the *cross ratio*. As applied in crystallography, it is usually called the *anharmonic ratio*, which is an unfortunate term because it includes, as a special case, the *harmonic ratio*. The cross ratio is based upon the segments of a straight line, known as a *transversal*, which is cut by a set of four rays issuing from a common *vertex* V. If the transversal is cut by the rays at points A, B, C, and D, not necessarily in that order, the mathematicians define the cross ratio as the ratio of the four directed segments of the straight line in the form

$$\text{cross ratio} \equiv \frac{AC}{CB} \bigg/ \frac{AD}{DB} .$$

This ratio has the important property of being invariant with any transversal. This invariance arises because the cross ratio depends only on the sines of the angles at the vertex which are intercepted by the segments, thus

$$\frac{AC}{CB} \bigg/ \frac{AD}{DB} = \frac{\sin AVC}{\sin CVB} \bigg/ \frac{\sin AVD}{\sin DVB} .$$

A corresponding invariance also holds for any straight line cut into segments by four coaxial planes.

The value of the cross ratio depends on the order of the lettering of the points. Since there are 24 permutations of four letters, there are 24 possible orders of the letters, but only six values of the resulting cross ratios are distinct, and these six are related.

Crystallographers use a cross ratio, which they call the "anharmonic ratio", and which they define in a somewhat different way. Specifically, for a series of points in the order $A'B'C'D'$ they define

$$\text{anharmonic ratio} \equiv \frac{A'B'}{B'C'} \bigg/ \frac{A'D'}{C'D'} .$$

This is the same as the mathematicians' cross ratio for the sequence $ACBD$, relettered as $A'B'C'D'$, but with a positive sign assigned to segment $A'D'$.

The cross ratio can be evaluated in terms of the indices of four tautozonal planes. If three of the four sets of the indices are known, the fourth set can be found, since the cross ratio can be experimentally determined, for example, by measuring the appropriate segments of the zone as delimited by the four poles. A set of formulas for cross ratio that are functions of the cotangents of the angles between the poles are also useful for indexing the poles.

Notes on history

Although it is not easy to find definite evidence in their writings, the cross ratio seems to have been known to the Greek geometers. Our present understanding of this subject is due to Möbius[1], who in 1827 set the pattern followed by present-day mathematical treatments[14,16,18]. Möbius called the ratio the *double-cut ratios* (*Doppelschnittverhältnisse*). Later, Chasles[3,6] called it the anharmonic ratio (*rapport anharmonique*), which is the term used by most crystallographers.

The cross ratio was not yet recognized in Miller's[4] 1839 textbook in crystallography, although a feebler sine relation was derived for his "sphere of projection". In an important paper[9] in 1859, however, he specifically noted that if four lines diverge from a point K and are intercepted by another line at points P, Q, R, and S, a product of certain segments of the last line is equal to the sines of the aperture angles of the corresponding segments in the following way.

$$\frac{\sin PKQ}{\sin RKQ} \cdot \frac{\sin RKS}{\sin PKS} = \frac{PQ}{RQ} \cdot \frac{RS}{PS}.$$

This can be manipulated to provide a cross ratio in the form

$$\frac{PQ}{RQ} \bigg/ \frac{PS}{RS} = \frac{\sin PKQ}{\sin RKQ} \bigg/ \frac{\sin PKS}{\sin RKS}.$$

The diagram on the left side is

$$P\overset{\frown}{Q}R\underset{\smile}{S}.$$

Crystallographers now commonly define their anharmonic ratio as

$$\frac{AB}{BC} \bigg/ \frac{AD}{CD},$$

whose diagram is

The sign of this is opposite to that of Miller. This is related to Möbius' double-cut ratio for the sequence *ACBD*, whose diagram is

Significant literature

1. August Ferdinand Möbius. Der barycentrische Calcul ein neues Hülfsmittel zur analytischen Behandlung der Geometrie dargestellt und insbesondere auf die Bildung neuer Classen von Aufgaben und die Entwickelung mehrerer Eigenschaften der Kegelschnitte. (Johann Ambrosius Barth, Leipzig, 1827) 453 pages and 77 figures, especially 243–265.
2. Carl Friedrich Gauss. [as an incidental part of a discussion of "Geometrische Seite der ternären Formen," recorded in Carl Friedrich Gauss Werke 2, Königlichen Gesellschaft der Wissenschaften zu Göttingen, 1863, 305–310] July, 1831.
3. M[ichel] Chasles. Aperçu historique sur l'origine et le développement des méthodes en géométrie, particulièrement de celles qui se rapportent à la géométrie moderne, suivi d'un mémoire de géométrie sur deux principes généraux de la science, la dualité et l'homographie. [Mémoires couronnés par l'Acad. royale des Sciences et Belles-lettres de Bruxelles (1837)] (M. Hayes, Bruxelles, 1837) 851 pages and diagrams. Second edition (Gauthier-Villars, Paris, 1875).
4. W. H. Miller. A treatise on crystallography. (J. & J. J. Deighton, Cambridge, 1839) 139 pages and 10 plates, especially 12–15.
5. Carl Georg Christian v. Staudt. Geometrie der Lage. (Bauer und Raspe, Nürnberg, 1847) 216 pages.
6. M[ichel] Chasles. Traité de géométrie supérieure. (Bachelier, Paris, 1852) 603 pages, 198 diagrams.
   Second edition (Gauthier-Villars, Paris, 1880) 585 pages, 198 diagrams; especially 1–53.
7. Karl Georg Christian v. Staudt. Beiträge zur Geometrie der Lage. (Fr. Korn, Nürnberg, 1856) 396 pages.
8. W. H. Miller. On the anharmonic ratio of radii normal to four faces of a crystal in one zone; and the change of axes of a crystal. *Phil. Mag.* **13** (1857) 96–103.
9. W. H. Miller. On the employment of the gnomonic projection of the sphere in crystallography. *Phil. Mag.* **18** (1859) 32–50, especially pages 32–43.
10. Rudolf Staudige. Lehrbuch der neueren Geometrie für höjere unterrichts-anstalten und zum selbststudien. (L. W. Seidel & Sohn, Wien, 1870) 365 pages, especially 4–45.
11. E. von Fedorow. Beiträge zur zonaler Krystallographie. 1. Ein besonderer Gang der zonalen Formenentwickelung. *Z. Kristallogr.* **32** (1900) 446–492, especially 456–457.
12. G. Cesàro. Démonstration simple de la loi de Miller. *Mineralog. Mag.* **17** (1916) 324–325.
13. E. Schiebold. Beiträge zur Auswertung der Laue-Diagramme. *Naturwis.* **10** (1922) 399–411, especially 406–407.

14. Levi S. Shively. An introduction to modern geometry. (Wiley, New York, 1939) 167 pages, especially 31–36, 103–107.
15. N. N. Padurow. Über Indizierung von Flächen und Kanten des Kristalles auf stereographischen Netz. *Neues Jahrb. f. Min., Monatshefte* (1949) 163–172.
16. Richard Courant and Herbert Robbins. What is mathematics? (Oxford University Press, New York, 1941) 521 pages, especially 172–178, 185.
17. J. D. H. Donnay. Démonstration de la relation des quatre faces en zone. *Bull. Soc. Royale Sciences Liège* **32** (1963) 909–911.
18. Howard Eves. A survey of geometry, Vol. 1. (Allyn and Bacon, Boston, 1963) 489 pages, especially 86–91, 98–99. Revised edition (1972), 442 pages, especially 73–77, 82–84.

Chapter 11

The indexing of Laue photographs

In Chapter 6 a general description of the identification of the symmetry directly from a conventional Laue photograph was given; in Chapter 8, identification of the symmetry through the study of the system of the nine zones in the projection was discussed. With this method, only planes with simple indices can be recognized. But these discussions are only a prelude to the treatment of the whole problem of interpreting a Laue photograph, because it may still be necessary to assign the appropriate index to each of the Laue spots. This can be done in a variety of ways. The easiest case occurs when a conventional Laue photograph has been taken with the primary x-ray beam parallel to a known zone axis. The most complicated case occurs when a Laue photograph recorded on a cylindrical film has been taken with the primary beam parallel to a general noncrystallographic direction. Any Laue photograph can be wholly indexed by using a general method based on the crystallographic principles of zone development. It seems better in a book like this, however, to start with the simplest cases and develop the general case later.

Indexing by zone–cone intersections

Some preliminary notes on the indexing of a Laue photograph were given in Chapter 5. Laue spots of tautozonal planes appear on the conventional Laue photograph on an ellipse that contains the record of the primary

267

beam. The index of any Laue spot can be deduced by identifying the two ellipses that intersect at that spot. The relation between planes and zones requires that the symbol of the plane can be deduced from the symbol of the two zones. The practical way of doing this is the following. If $[u_1v_1w_1]$ and $[u_2v_2w_2]$ are the indices of the two zones that intersect at a Laue spot, then the relations of (6), (7), and (8) of Chapter 8 hold for the three sets of indices, specifically

(6): $\qquad hu_1 + kv_1 + lw_1 = 0,$

(7): $\qquad hu_2 + kv_2 + lw_2 = 0,$

(8): $\qquad \dfrac{h}{v_1w_2 - w_1v_2} = \dfrac{k}{w_1u_2 - u_1w_2} = \dfrac{l}{u_1v_2 + v_1u_2}.$

When h, k, and l have no common factor, then

$$h = v_1w_2 - w_1v_2,$$
$$k = w_1u_2 - u_1w_2, \qquad\qquad (1)$$
$$l = u_1v_2 - v_1u_2.$$

A rule of thumb for determining h, k, and l is to write down the sequences u_1, v_1, w_1 and u_2, v_2, w_2 twice and treat the middle part like a determinant as follows.

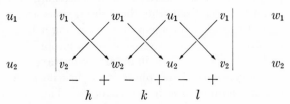

This procedure gives the value of h, k, and l as in (1). Care must be taken to find the true signs of the indices of the plane so obtained. Let us find the symbol of the plane at the intersection of the zones $[001]$ and $[\bar{1}10]$. According to the rule, we write

$$0 \begin{vmatrix} 0 & 1 & 0 & 0 \end{vmatrix} 1$$
$$\bar{1} \begin{vmatrix} 1 & 0 & \bar{1} & 1 \end{vmatrix} 0,$$

and the symbol $(\bar{1}\bar{1}0)$ is obtained. However, it is also legitimate to write

$$\bar{1} \begin{vmatrix} 1 & 0 & \bar{1} & 1 \end{vmatrix} 0$$
$$0 \begin{vmatrix} 0 & 1 & 0 & 0 \end{vmatrix} 1,$$

from which (110) is obtained. In order to select the true signs, the correct position of the pole in the zone must be ascertained.

The method of zone–cone intersection is useful for indexing Laue spots of crystals with high symmetry. Let us assume a cubic crystal with the x-ray beam directed along [001], as illustrated in Fig. 1. In this case [100] and [010] are perpendicular to the x-ray beam and parallel to the photographic film. The intersections of these zones with the photographic film are Z_1 and Z_2, respectively, which are located on two perpendicular straight lines OZ_1 and OZ_2, that radiate from the origin O. Any zone [$u0w$] gives rise to a cone whose intersection with the film is an ellipse with its long axis along OZ_1, the direction of [$\bar{1}00$], while any zone [$0vw$] gives rise to a cone whose ellipse has a long axis along OZ_2, the direction of [010].

A set of ellipses can be drawn for the different indices of the zones for the given crystal-to-plate distance. For this method, Schiebold[1] developed the chart that is illustrated in Fig. 2. In this diagram ellipses for constant values of l and different values of h and k are given. With the aid of a chart like Fig. 2, the symbols of the Laue spots can be easily identified. Using the method just discussed, one finds that the intersection of the two zones determines the reflection from ($\bar{3}\bar{6}2$).

The drawing of the ellipses may be tedious. It is easier to use the stereographic projection in which the ellipses of the Laue photograph are transformed into circles. The stereographic projection of the zone circles cannot

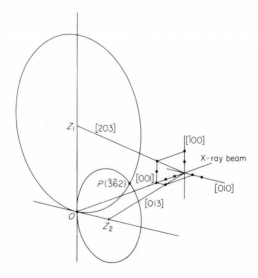

Fig. 1. Intersection of zone ellipses in a conventional Laue photograph.

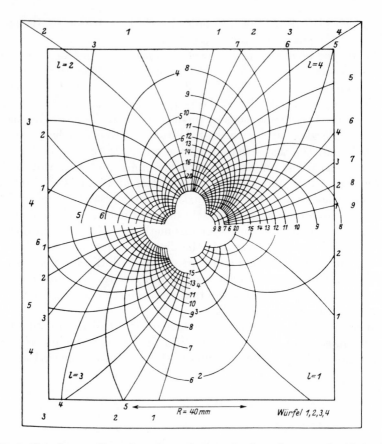

Fig. 2. Net of zone ellipses in a Laue photograph of a cubic crystal with incident beam parallel to [001]. In each quadrant one ellipse corresponds to constant values of l and different values of h and k. [After Schiebold[6], p. 38, Fig. 15.]

be used directly on the Laue photograph but the Laue spots can be easily identified with the corresponding intersections of the zone circles in the projection.

In order to draw the zone circles in the stereographic projection, we can make use of the facts that every zone circle contains the pole of (001) and that the center of the circle of a zone [uvw] is the projection of the zone axis; the radius of the circle is the gnomonic distance of this point from the center of the gnomonic projection. If the direction of the x-ray beam is a crystallographic direction [$u_0v_0w_0$], the angle ψ that it forms with any other zone [uvw], namely

$$\psi = [u_0v_0w_0] \wedge [uvw], \tag{2}$$

is given by

$$\cos \psi = \frac{\sum u_0 u a^2 + \sum (u_0 v + v_0 u) ab \cos \gamma}{L(u_0 v_0 w_0) L(uvw)} \tag{3}$$

where

$$\sum u_0 u a^2 = u_0 u a^2 + v_0 v b^2 + w_0 w c^2, \tag{4}$$

$$\sum (u_0 v + v_0 u) ab \cos \psi = (u_0 v + v_0 u) ab \cos \gamma$$
$$+ (v_0 w + w_0 v) bc \cos \alpha$$
$$+ (w_0 u + u_0 w) ac \cos \beta, \tag{5}$$

and $L(u_0 v_0 w_0)$ and $L(uvw)$ are the parameters of the rows $[u_0 v_0 w_0]$ and $[uvw]$, that is, the lengths of the respective lattice vectors. The parameter of a row is given by

$$L(uvw) = (u^2 a^2 + v^2 b^2 + w^2 c^2$$
$$+ 2vwbc \cos \alpha + 2wuca \cos \beta + 2uvab \cos \gamma)^{.1/2} \tag{6}$$

The ψ angle in (3) is also given by

$$\tan \psi = \frac{(\sum A + 2 \sum B)^{1/2}}{V^*(\sum C + \sum D)} \tag{7}$$

where

$$\sum A = (v_0 w - w_0 v)^2 a^{*2} + (u_0 w - w_0 u)^2 b^{*2} + (u_0 v - v_0 u)^2 c^{*2},$$

$$\sum B = (w_0 u - u_0 w)(u_0 v - v_0 u) b^* c^* \cos \alpha^*$$
$$+ (u_0 v - v_0 u)(v_0 w - vw_0) c^* a^* \cos \beta^* \tag{8}$$
$$+ (v_0 w - w_0 v)(w_0 u - wu_0) a^* b^* \cos \gamma^*,$$

$$\sum C = u_0 u a^2 + v_0 v b^2 + w_0 w c^2,$$

$$\sum D = (v_0 w + w_0 v) bc \cos \alpha + (w_0 u + u_0 w) ca \cos \beta + (u_0 v + v_0 u) ab \cos \gamma.$$

Equations (3) and (8) are of general application. The increase of symmetry involves important simplifications in the explicit form of the equations. Table 1 collects the equations for $\cos \psi$ for various crystallographic systems, except triclinic, for which the general equations given here apply.

In these equations, $[u_0 v_0 w_0]$ is the row parallel to the x-ray beam. If the x-ray beam is along an axial direction, the system of zone circles that defines most of the Laue spots can be easily drawn. In the case of a conventional Laue photograph of a cubic crystal taken with the beam along [001], (7) is highly simplified. For instance, for zones of the type $[u0w]$,

Table 1

Angle ψ between two lattice rows $[u_0v_0w_0]$ and $[uvw]$ for crystals other than triclinic

Monoclinic

$$\cos\psi = \frac{u_0ua^2 + v_0vb^2 + w_0wc^2 + (w_0u + u_0w)ca\cos\beta}{L_{u_0v_0w_0}L_{uvw}},$$

where $L = (u^2a^2 + v^2b^2 + w^2c^2 + 2wuca\cos\beta)^{1/2}.$

Orthorhombic

$$\cos\psi = \frac{u_0ua^2 + v_0vb^2 + w_0wc^2}{L_{u_0v_0w_0}L_{uvw}},$$

where $L_{uvw} = (u^2a^2 + v^2b^2 + w^2c^2)^{1/2}.$

Tetragonal

$$\cos\psi = \frac{(u_0u + v_0v)a^2 + w_0wc^2}{L_{u_0v_0w_0}L_{uvw}},$$

where $L_{uvw} = [(u^2 + v^2)a^2 + w^2c^2]^{1/2}.$

Hexagonal

$$\cos\psi = \frac{[u_0u + v_0v + \frac{1}{2}(u_0v + v_0u)]a^2 + w_0wc^2}{L_{u_0v_0w_0}L_{uvw}},$$

where $L_{uvw} = [(u^2 + v^2 + uv)a^2 + w^2c^2]^{1/2}.$

Rhombohedral

$$\cos\psi = \frac{\{u_0u + v_0v + w_0w + [u_0(v + w) + v_0(w + u) + w_0(u + v)]\cos\alpha\}a^2}{L_{u_0v_0w_0}L_{uvw}},$$

where $L_{uvw} = a[(u^2 + v^2 + w^2) + 2(vw + wu + uv)\cos\alpha]^{1/2}.$

Cubic

$$\cos\psi = \frac{u_0u + v_0v + w_0w}{L_0L},$$

where $L_{uvw} = a(u^2 + v^2 + w^2)^{1/2}.$

(7) reduces to

$$\tan\psi = \frac{u}{w} \tag{9}$$

and for zones of the type $[0vw]$, it reduces to

$$\tan\psi = \frac{v}{w}. \tag{10}$$

The centers of the corresponding zone circles are along two perpendicular directions; their distances to the center of the projection are $R \tan \frac{1}{2}\psi$ where R is the radius of the reference sphere that can be taken as the crystal-to-plate distance. A stereographic projection of a set of zone circles for a cubic crystal is given in Fig. 3. In this symmetry the identification of the Laue spots in the photograph is an easy task.

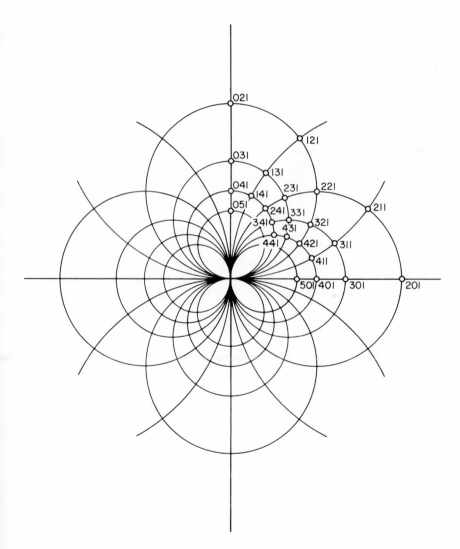

Fig. 3. Stereographic projection of zone circles of a cubic crystal.

In normal practice, however, the axis of the crystal is not so restrictively oriented, and the resulting figure is complicated. The method, however, is useful in giving a pictorial way of understanding of the general relation of Laue spots to the intersections of diffraction cones according to the views of von Laue. This appearance is shown in any kind of Laue photograph.

Indexing through the gnomonic projection

General principles. The indexing of a Laue photograph can also be done in a very advantageous way by using the gnomonic projection. The simplest case, of course, occurs when the orientation of the crystal is such that the normal pattern of the gnomonic projection is obtained. The normal pattern corresponds to the same orientation as the normal pattern of the stereographic projection.

As was shown in Chapter 3, the main characteristic of the gnomonic projection is that the zones project as general straight lines. The two axial zones [100] and [010] determine two important zone lines in the normal pattern. Other zones not belonging to the axial category are known as radial zone lines, of which the most important is that containing the poles of the planes (001) and (110). This zone is the main radial zone line. The section of the gnomonic projection that contains the nine principal poles (001), (101), (011), ($\bar{1}$01), (0$\bar{1}$1), (111), ($\bar{1}$11), ($\bar{1}\bar{1}$1), and (1$\bar{1}$1) is known as the primitive gnomogram (Fig. 4). The poles of the primitive gnomogram determine a plane lattice in that projection, whose translations are p_0 and q_0; these are inclined at an angle γ^* to each other. The translation p_0 is determined by the distance between the poles (001) and (101), while q_0

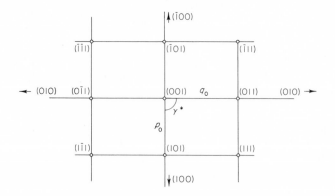

Fig. 4. Primitive gnomogram.

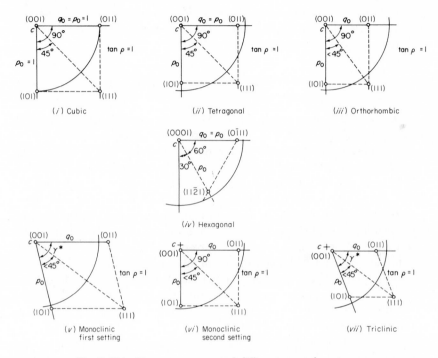

Fig. 5. Primitive gnomograms of different crystal systems.

is determined by the distance between poles (001) and (011). The co-ordinates of the pole (111) in the reference system of the two axial zones [100] and [010] are obviously p_0 and q_0. These three constants, as well as x_0, y_0, the coordinates of the pole (001) in the normal pattern of the gnomonic projection, are different for each crystallographic system, as shown in Fig. 5, and serve to identify it. The gnomonic projection of a cubic crystal, Fig. 5i, is based on a square network whose sides are equal to the distance of the photographic plate from the crystal; this distance is taken as the radius of the reference sphere used for the projection. In general, the lengths of the sides for a cubic crystal are equal to the radius of the sphere.

If the crystal is tetragonal, the normal pattern of the gnomonic projection is also a square network, as shown in Fig. 5ii. However, this network differs from the one of the cubic crystal in that the sides of the squares are generally equal not to the radius of the reference sphere, but to this distance multiplied by the axial ratio c/a of the crystal.

If the crystal is orthorhombic, the normal pattern of the gnomonic projection shows a rectangular network, as in Fig. 5*iii*. The ratio of the lengths of the sides of the rectangular net is the reciprocal of the corresponding axial ratios in the crystal cell:

$$p_0 = R\frac{c}{a}, \qquad q_0 = R\frac{c}{b}. \tag{11}$$

If the crystal is hexagonal, the network is hexagonal, with $\gamma^* = 60°$, as shown in Fig. 5*iv*. The lengths p_0, q_0 are equal to Rc.

If the crystal is monoclinic, the normal pattern depends on the setting. There are two different accepted settings for a monoclinic crystal. In what is known as the first setting, the unique axis is labeled c, and [001] is placed perpendicular to the plane of projection, while the pole (001) coincides with the center of the projection, as shown in Fig. 5*v*. In this case, the network is oblique, the angle γ^* is not specialized, and the translations p_0 and q_0 are proportional to c/a and c/b, respectively. In the second setting, the unique axis is labeled b and placed in an east–west direction, as in Fig. 5*vi*. Then [001] is perpendicular to the plane of projection and (001) is below the center of the projection. The distance from (001) to the center of the gnomonic projection is equal to $R\tan(90° - \beta^*)$. The network is rectangular with sides proportional to c/a and c/b, respectively.

If the crystal is triclinic, the network is oblique and its origin is off the center of the projection, as in Fig. 5*vii*. The angle of the parallelogram is γ^* and the sides are proportional to c/a and c/b, respectively.

The parameters p_0 and q_0, determined by the distance between the poles (001) and (101) and between (001) and (011), have the important characteristic of being proportional to c/a and c/b, respectively. Any other pole (hkl) is determined by two coordinates p, q on the same reference system of the two axial zones [100] and [010], such that

$$p = \frac{h}{l}p_0 \qquad \text{and} \qquad q = \frac{k}{l}q_0, \tag{12}$$

from which the indices of the pole can be determined. This procedure can be applied to any kind of crystal, provided that due consideration is given to the actual symmetry of the crystal and to the angular relations between zones. The values p_0, q_0 always depend on the axial ratio of the given crystal.

Some general precautions must be taken into account when indexing the gnomonic projection of the Laue photograph. First, it is difficult to determine with great accuracy the distance from the center of the Laue photograph to a Laue spot that has a small angle of diffraction. Reference to Fig. 8 of Chapter 5 will show that planes which have a small angle of diffraction appear as poles that lie in the outer regions of the gnomonic projection. As a result, poles lying in the outer region of the projection do not fit exactly on the points of the gnomonic net based on the fundamental parameters p_0 and q_0, which are determined by poles near the center of the projection. Small adjustments for outer poles may be needed to move them to the nearest net points on the fundamental net. With a little practice, it is seen that only very minor corrections are, indeed, needed. As shown in Chapter 5, the Laue spots that lie in the region near the record of the direct x-ray beam correspond to points of the first level of the reciprocal-lattice stack whose planes are perpendicular, or almost perpendicular, to the x-ray beam, and therefore the lattice defined by such poles is just an enlarged replica of the mesh of that reciprocal-lattice level. Some of the points nearer the origin of the gnomogram correspond to the second level, as discussed in detail in Chapter 5. They are located nearer the center of the mesh defined by the outermost poles. Small adjustments are again needed to bring them to the right position. Other poles that are not on those specialized positions correspond to the reciprocal-lattice points of the third, fourth, etc., levels of the same stack of reciprocal-lattice planes. Their positions can easily be deduced because the coordinates of the poles are p_0/n or q_0/n where n is the level number of the corresponding plane in the reciprocal lattice.

In order to determine the index of a pole in the gnomogram, the following rule should be applied. First, a net with equidistant parallel zone lines must be drawn. In the special setting that gives the normal pattern, the recognition of the different poles presents no great difficulty. Even in a monoclinic or triclinic crystal, the pole of (001), which is a very important crystal plane, can be recognized by the great number of radial zones that can be traced through it.

Once the network system has been established and the poles (001), (011), and (101) recognized, the two coordinates p, q of the other poles can be easily read. The first corresponds to the value h', the second to k'. The provisional indices $h'k'l'$ are thus obtained for each pole. The true indices hkl are deduced from $h'k'l'$ by multiplying these three numbers by the same factor, so that they are transformed into three integers with no common factor. By proceeding in this way, indices can be given to every pole in the projection. In the next section, an example of the use of this method in the indexing of a Laue photograph of a cubic crystal is given.

Indexing a conventional Laue photograph of a cubic crystal. The poles of the Laue spots in the gnomonic projection are always found to lie on two sets of parallel lines; the main lines are those that correspond to the axial zones [100] and [010], and both sets are mutually perpendicular.

These lines can then be labeled in terms of the p, q values, namely

$$\frac{h}{l} = \ldots, 4, 3, 2, \tfrac{3}{2}, \tfrac{4}{3}, 1, \tfrac{3}{4}, \tfrac{2}{3}, \tfrac{1}{2}, \ldots, 0, \ldots, -\tfrac{1}{2}, -\tfrac{2}{3}, -\tfrac{3}{4}, -1, \ldots,$$

$$\frac{k}{l} = \ldots, 4, 3, 2, \tfrac{3}{2}, \ldots, 0, -1, -\tfrac{4}{3}, -\tfrac{3}{2}, -2, \ldots \tag{13}$$

and in general $h/l = p/q$ and $k/l = q/p$. It follows from what has been said that p and q are, in general, small positive or negative integers. From the intersections h/l and k/l the index of the spot can be deduced. The unit of spacing p_0 and q_0 of the lattice defined in the gnomonic projection is R_g', that is, the radius of the gnomonic projection.

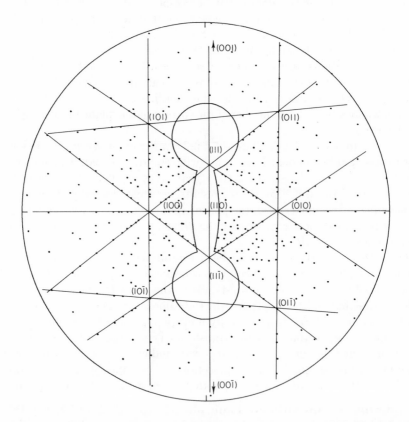

Fig. 6. Gnomonic projection of a cylindrical Laue photograph of NaCl (projection boundary at $\rho = 71.0°$).

In general, the x-ray beam is not incident along [001], but it may be along any direction [uvw]. In this more general case, the gnomonic projection of the cubic crystal is not necessarily symmetrical, but the Laue photograph can also be indexed by using a generalization of the procedure just discussed, as suggested by Lonsdale[8]. Two systems of nearly parallel rows can be recognized in the gnomonic projection. An example is shown in Fig. 6. Each system is a set of rows parallel to a pseudo-main row passing through the origin. Pseudo-main rows correspond, in the present case, to relatively simple rows perpendicular to the incidence direction [uvw], for instance, [$0w\bar{v}$], [$\bar{w}0u$], [$v\bar{u}0$]. Each pair of these pseudo-main zones makes angles that can be easily recognized. The expected values can be computed, for instance, by

$$\cos [0w\bar{v}] \wedge [\bar{w}0u] = -\frac{uv}{(w^2 + v^2)(u^2 + w^2)}, \qquad (14)$$

$$\cos [\bar{w}0u] \wedge [v\bar{u}0] = -\frac{vw}{(u^2 + w^2)(v^2 + u^2)} \qquad (15)$$

and so on.

The poles of the planes (hkl) belonging to these zones are on sets of parallel straight lines whose indices are related by

$$\frac{kw - lv}{hu + kv + lw} = \frac{p'}{q'},$$

$$\frac{-hw + lu}{hu + kv + lw} = \frac{p'}{q'},$$

$$\frac{hv - ku}{hu + kv + lw} = \frac{p'}{q'} \qquad (16)$$

where p' and q' are again small positive or negative integers.

In the cubic crystal there are many rows mutually perpendicular. In the set of rows perpendicular to the direction of incidence [uvw] there is one perpendicular also to [$0w\bar{v}$], namely [$u'v'w'$]. The condition for a row [uvw] to be perpendicular to [$u'v'w'$] is that

$$uu' + vv' + ww' = 0, \qquad (17)$$

so that

$$0u' + wv' - vw' = 0 \qquad (18)$$

also holds in this case. The new direction $[u'v'w']$ has indices

$$u' = -(v^2 + w^2)/u,$$

$$v' = v,$$

$$w' = w.$$

The two rows determine a new rectangular reference system which facilitates the indexing of the poles in the gnomonic projection.

In general, any two rows passing through the origin of the gnomonic projection determine two sets of parallel rows, for which

$$\frac{hu' + kv' + lw'}{hu + hv + lw} = \frac{h}{q} \tag{19}$$

where, of course, (15) holds also. In this case, the spacing determined by the integer p/q is given by

$$D' = R_g\left(\frac{u^2 + v^2 + w^2}{u'^2 + v'^2 + w'^2}\right)^{1/2}. \tag{20}$$

This can be used to determine in an easy way the value p/q for all sets of lines. The indices (hkl) of a given pole can be deduced by combining and solving the equations of the two lines that intersect in the pole.

Indexing through stereographic projection

Advantages. Normal patterns of specialized orientations are not a universal feature of the projection of a Laue photograph; usually the researcher is confronted with the necessity of interpretating a projection in general orientation. The gnomonic projection is then distorted and its interpretation is not straightforward. Even in the case of general incidence, the indices of the Laue spots can be determined with relative simplicity only in cases when the x-ray beam is incident along a known direction. All the methods described in the preceding sections are based on the knowledge of the direction of incidence. Fortunately, this special incidence is not needed to index the poles in a stereographic projection, and as a matter of fact, it does not introduce any important simplification in the procedure.

The stereographic projection (Chapter 2) is simple to use and therefore is most appropriate for solving completely the problem of indexing. Zones are easily recognized and the poles can be indexed by zone intersection. The problem of indexing is a simple crystallographic one that can be solved in many different ways. However, three fundamental methods are commonly used. The parametric relations of a given pole may be used; the researcher may make use of the relations between four tautozonal planes and their indices; or he may apply the more sophisticated method of indices to be expected in zone development. In all these methods lengthy calculations are used, and the indexing of the whole photograph can be very time-consuming. The whole problem can, however, be solved in a more direct way by having at hand a set of angles between poles of known indices for a given crystal. Finally, the indices can be determined simply by visual comparison of the stereographic projection with standard projections along different orientations.

Use of interplanar angles. Indexing of the projections of a Laue photograph can be accomplished by comparing the angles between the poles with tabulated values for standard cases. It is self-evident that a proper selection of indices of the poles in a zone implies a proper sequence of angles between the poles. This is the basic idea of the procedures described in the preceding section.

If we recognize the fact that a projection of a cubic crystal is independent of the crystal species, we are led to the corollary that a set of interfacial angles is just what is needed for a rapid identification of the poles in the projection. The practical application of the idea has been possible through the use of tables of angles between planes. The interplanar angle ψ between $(h_1k_1l_1)$ and $(h_2k_2l_2)$ of a cubic crystal can easily be calculated by

$$\cos \psi = \frac{h_1h_2 + k_1k_2 + l_1l_2}{(h_1^2 + k_1^2 + l_1^2)(h_2^2 + k_2^2 + l_2^2)} . \tag{21}$$

A set of angles up to {321} for cubic crystals was calculated by Bozorth[2] and the series was extended up to {554} by Peavler and Lenusky[13]. Also, a set of angles up to {531} is given by J. D. H. Donnay and Gabrielle Donnay[14]. These tables are very convenient for most orientation work.

Tables of interplanar angles for other crystal systems can also be calculated, although in this case the calculations must be made using the specific crystallographic translations for the specific crystal being considered. The most convenient case is the tetragonal one, where the angles depend only on the parameters a and c. In this case,

$$\cos \psi = \frac{\left(\dfrac{h_1h_2 + k_1k_2}{a^2}\right) + \dfrac{l_1l_2}{c^2}}{\left[\left(\dfrac{h_1^2 + k_1^2}{a^2} + \dfrac{l_1^2}{c^2}\right)\left(\dfrac{h_2^2 + k_2^2}{a^2} + \dfrac{l_2^2}{c^2}\right)\right]^{1/2}} . \tag{22}$$

A set of interplanar angles as a function of c/a ranging from 0.500 to 0.700 has been calculated by Frounfelker and Hirthe[15].

In the case of hexagonal crystals, the calculation is further complicated by the nonorthogonality of a and b. In this case the relation is

$$\cos \psi = \frac{\dfrac{2}{3a^2}(2h_1h_2 + 2k_1h_2 + h_1k_2 + k_1h_2) + \dfrac{l_1l_2}{c^2}}{\left[\left(\dfrac{4}{3a^2}(h_1^2 + k_1^2 + h_1k_1) + \dfrac{l_1^2}{c^2}\right)\left(\dfrac{4}{3a^2}(h_2^2 + k_2^2 + h_2k_2) + \dfrac{l_2^2}{c^2}\right)\right]^{1/2}}.$$

(23)

A set of angles has been calculated by Taylor and Leber[11] for c/a ranging from 1.500 to 2.000. The same authors give some angles for Be $(c/a = 1.5847)$, Ti $(c/a = 1.5873)$, Zr $(c/a = 1.5893)$, Mg $(c/a = 1.6235)$, Zn $(c/a = 1.8563)$, and Cd $(c/a = 1.8859)$.

Orthorhombic crystals cannot be treated except for specific cases. The ψ angle is given by

$$\cos \psi = \frac{\dfrac{h_1h_2}{a^2} + \dfrac{k_1k_2}{b^2} + \dfrac{l_1l_2}{c^2}}{\left[\left(\dfrac{h_1^2}{a^2} + \dfrac{k_1^2}{b^2} + \dfrac{l_1^2}{c^2}\right)\left(\dfrac{h_2^2}{a^2} + \dfrac{k_2^2}{b^2} + \dfrac{l_2^2}{c^2}\right)\right]^{1/2}}.$$

(24)

Sets of angles for gallium,[16] α uranium,[10] bismuth, and antimony[12] have been published.

If routine work of orientation is required, a set of Laue photographs of a crystal in known orientations can be used as a reference file. This procedure has been applied to face-centered and body-centered cubic crystals by Majima and Togino[4,5] and by Dunn and Martin[9]. The method can be useful only if a large set of Laue photographs is available. Majima and Togino use 55 standard Laue photographs, and Dunn and Martin, 75. The whole set of Dunn and Martin covers some 300 patterns. For other crystal systems the method is impractical.

Determination of crystal orientation: standard projections

One of the practical uses of the Laue method is to determine crystal orientation. Of course, crystal orientation is known when the projection of the poles of the Laue spots has been indexed. However, the researcher whose aim is just to determine the orientation of the crystal does not have to go through the whole process of indexing all the poles; he need only recognize the symmetry pattern to determine the angle that the important poles make with the x-ray beam, the center of the projection. A method ordinarily used for this purpose is due to Schiebold and Sachs[3], based on the use of a conventional transmission Laue photograph. From the Laue spots a stereographic projection is prepared. The projection is not, in general, symmetric. The orientation can be determined by identifying an important zone with the aid of a standard projection. A rotation of the projection, using the Wulff net, is ordinarily needed to make the important zone coincide with the corresponding one on the standard projection. Rotation can be avoided if standard projections in desired crystal orientations are prepared. With these available, important zones can easily be recognized, thus diminishing the probability of introducing errors. A method of preparing standard stereographic projections is through the calculation of the coordinates of the poles in the projection[7,20]. With this method any desired degree of accuracy can be obtained, so that the precision of the standard projection is limited only by the plotting of the calculated distances, which need not exceed a few tenths of a degree.

For a cubic crystal the x, y stereographic coordinates can easily be com-

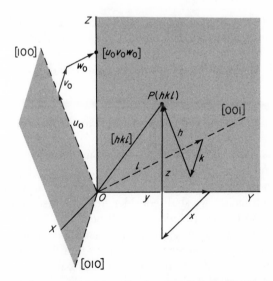

Fig. 7. Standard Cartesian system and cubic axes discussed in the text. The stereographic coordinates x_s and y_s of the pole $P(hkl)$ are proportional to x and y, respectively.

puted for every desired (hkl) and for a projection plane in a given orientation. Let us choose an orthogonal Cartesian system, and let OXY be the plane of projection, and O its center. In addition, let us orient the cubic crystal in such a way (Fig. 7) that $[001]$ lies in the YZ plane, and therefore, the X axis lies in the plane O, $[100]$, $[010]$. In this way the pole (001) is always located on the positive side of the Y axis of the projection in a manner similar to that chosen in the standard projection given by Schiebold and Sachs[3].

For the cubic crystal, one can choose a direction $[u_0v_0w_0]$ for which the standard projection has to be computed. Then the pole $(u_0v_0w_0)$ lies in the center of the projection. The crystal direction $[u_0v_0w_0]$ coincides with OZ in this new system and the crystallographic indices (hkl) can easily be transformed into the (HKL) of the new system. Then the x_s, y_s coordinates of every pole (hkl) can easily be computed in terms of

$$x_s = \frac{(u_0k - v_0h)C}{[C(h^2 + k^2 + l^2)^{1/2} + u_0h + v_0k + w_0l](u_0{}^2 + v_0{}^2)^{1/2}}R_s,$$

$$y_s = \frac{u_0(u_0l - w_0h) + v_0(v_0l - w_0k)}{[C(h^2 + k^2 + l^2)^{1/2} + u_0h + v_0k + w_0l](u_0{}^2 + v_0{}^2)^{1/2}}R_s,$$

(25)

in which

$$C = (u_0{}^2 + v_0{}^2 + w_0{}^2)^{1/2}.$$

If the projection is to be compared with a given Laue photograph, it is a matter of routine to generate the set of indices (hkl) that lie inside a limiting sphere having a radius that may be a function of the γ_{min} of the x-ray radiation used. Then (25) is used for every (hkl) to obtain the desired projection.

Direct indexing of the Laue spots

The relations that exist between the gnomonic projection of the pole and the Laue spot on the photograph allow one to index the Laue spot directly from its x, y coordinates in the film by a method due to Ahmed and Georgeoura[18]. For simplicity, we shall consider first the case of a cubic crystal; then the extension to the triclinic case will be given. The method outlined here permits indexing a Laue spot from measurements of the x, y coordinates. The radius of the film, or the crystal-to-film distance, is to be given, as well as the incidence direction $[u_0v_0w_0]$ of the x-ray beam and the direction $[u_1v_1w_1]$ of the crystal along the camera axis.

Cubic case. In a cubic crystal there are several sets of lattice rows that are mutually perpendicular; let us assume that $[u_0v_0w_0]$, $[u_1v_1w_2]$, and $[u_2v_2w_2]$ constitute one of those sets. Let us define also an orthogonal reference system in the experiment such that

Z is along the incident x-ray beam and coincides with $[u_0v_0w_0]$;
Y is along the camera axis and coincides with $[u_1v_1w_1]$;
X, perpendicular to Y and Z, is defined as $[u_2v_2w_2]$, where

$$u_2 = v_1w_0 - v_0w_1,$$

$$v_2 = w_1u_0 - w_0u_1,$$

$$w_2 = u_1v_0 - u_0v_1. \qquad (26)$$

In the flat transmission Laue case the equation of the film plane, which coincides with the plane of the gnomonic projection, is given by

$$u_0x + v_0y + w_0z = D(u_0{}^2 + v_0{}^2 + w_0{}^2)^{1/2} \qquad (27)$$

where D is the crystal-to-film distance. The normal to a crystal plane (hkl) is given by

$$\frac{x}{h} + \frac{y}{k} + \frac{z}{l}. \qquad (28)$$

The intersection of this normal with the plane (27) is the gnomonic pole (hkl).

Let us define a reference system in the film parallel to the one first described, such that the origin is the point of incidence of the primary x-ray beam on the film. In this case, any point on the film has $z = 0$. The x, y coordinates of the gnomonic pole are given by

$$x_g = \frac{u_2h + v_2k + w_2l}{u_0h + v_0k + w_0l} \frac{DC}{A}, \qquad (29)$$

$$y_g = \frac{u_1h + v_1k + w_1l}{u_0h + v_0k + w_0l} \frac{DC}{B} \qquad (30)$$

where

$$C = (u_0{}^2 + v_0{}^2 + w_0{}^2)^{1/2}, \qquad (31)$$

$$B = (u_1{}^2 + v_1{}^2 + w_1{}^2)^{1/2}, \qquad (32)$$

$$A = (u_2{}^2 + v_2{}^2 + w_2{}^2)^{1/2}. \qquad (33)$$

The coordinates x, y of the Laue spot on the film are related to the coordinates of the pole in a gnomonic projection by

$$x = \frac{x_g}{r_g} (x^2 + y^2)^{1/2}, \tag{34}$$

$$y = \frac{y_g}{r_g} (x^2 + y^2)^{1/2}, \tag{35}$$

where r_g is the distance of the gnomonic pole to the center projection. Dividing (34) by (35) and introducing the values of (29) and (30), one gets

$$\frac{y}{x} = -\frac{B}{A} \frac{u_2 h + v_2 k + w_2 l}{u_1 h + v_1 k + w_1 l}, \tag{36}$$

$$\frac{y}{(x^2 + y^2)^{1/2}} = \frac{DC}{A} \frac{u_1 h + v_1 k + w_1 l}{u_0 h + v_0 k + w_0 l}. \tag{37}$$

These two equations have three unknowns h, k, and l. By introducing $p = h/l$ and $q = k/l$, the following expressions can be obtained.

$$\frac{u_2 p + v_2 q + w_2}{u_1 p + v_1 q + w_1} = -\frac{A}{B} \frac{x}{y},$$

$$\frac{u_1 p + v_1 q + w_1}{u_0 p + v_0 q + w_0} = -\frac{B}{DC} \frac{y \rho_g}{(x^2 + y^2)^{1/2}}, \tag{38}$$

in which the negative signs are introduced to conform with the orientation of setting 1, Chapter 9.

From (38) two equations with two unknowns p, q are explicitly obtained, namely

$$p\left(u_2 + \frac{u_1 A}{B} \frac{x}{y}\right) + q\left(v_2 + \frac{v_1 A}{B} \frac{x}{y}\right) = -\left(\frac{w_1 A}{B} \frac{x}{y} + w_2\right),$$

$$p\left[u_1 - \frac{u_0 B}{D \cdot C} \frac{y \rho_g}{(x_2 + y_2)^{1/2}}\right] + q\left[v_1 - \frac{v_0 B}{D \cdot C} \frac{y \rho_g}{(x^2 + y^2)^{1/2}}\right] =$$

$$\frac{w_0 B}{D \cdot C} \frac{y \rho_g}{(x^2 + y^2)^{1/2}} w_0 - w_1. \tag{39}$$

For simplicity, let us call

$$E = u_2 + \frac{u_1 A}{B} \frac{x}{y},$$

$$F = v_2 + \frac{v_1 A}{B} \frac{x}{y},$$

$$G = u_1 + \frac{u_0 B}{D \cdot C} \frac{y \rho_g}{(x^2 + y^2)^{1/2}},$$

$$I = v_1 - \frac{v_0 B}{D \cdot C} \frac{y \rho_g}{(x^2 + y^2)^{1/2}}, \tag{40}$$

$$J = - \left(\frac{w_1 A}{B} \frac{x}{y} + w_2 \right),$$

$$M = \frac{w_0 B}{D \cdot C} \frac{y \rho_g}{(x^2 + y^2)^{1/2}} w_0 - w_1.$$

The unknowns p and q can easily be obtained by solving

$$p = \frac{\begin{vmatrix} J & F \\ M & I \end{vmatrix}}{\begin{vmatrix} E & F \\ G & I \end{vmatrix}}, \qquad q = \frac{\begin{vmatrix} E & J \\ G & M \end{vmatrix}}{\begin{vmatrix} E & F \\ G & I \end{vmatrix}}. \tag{41}$$

Thus it is possible to index a Laue spot $L(p, q)$ from its film coordinates $L(x, y)$ for a transmission flat Laue photograph of a cubic crystal, in the particular conditions studied.

In x-ray diffuse-scattering studies using the Laue method, it is customary to take a set of Laue photographs with the crystal mounted along a crystal axis, as discussed in Chapter 12. The x-ray beam is initially $[u_0 v_0 w_0]$; successive photographs are taken at constant intervals from the initial orientation. One can easily index a Laue photograph where the direction of incidence $[u'v'w']$ makes an angle ϕ with $[u_0 v_0 w_0]$: since u', v', w' are direction cosines in the cubic system, we have

$$\cos \phi = \frac{u'u_0 + v'v_0 + w'w_0}{u_0^2 + v_0^2 + w_0^2}. \tag{42}$$

The direction $[u'v'w']$ is also perpendicular to the camera axis $[u_1 v_1 w_1]$.

Therefore,

$$u'u_1 + v'v_1 + w'w_1 = 0, \tag{43}$$

$$u'^2 + v'^2 + w'^2 = 1. \tag{44}$$

Since angle ϕ is known, the symbol $[u'v'w']$ can be computed. These values can be substituted for u, v, w, in (40) and (41); in this way, any transmission Laue photograph of a cubic crystal can be indexed provided that the incident x-ray beam makes an angle ϕ with a given known crystallographic direction $[u_0v_0w_0]$ and that this direction is perpendicular to the direction $[u_1v_1w_1]$ along which the crystal is mounted.

For back-reflection Laue photographs, one can also use (40) and (41) to obtain the p, q corresponding to a Laue spot with coordinates x, y. In this case the sign of x is changed.

For cylindrical Laue photographs it is convenient first to transfer the two coordinates x_c, y_c, which are actually read on the cylindrical film, to coordinates x_t, y_t, which would have been obtained from a flat transmission (or back-reflection) film. The relationships between x_t, y_t and x_c, y_c were obtained in (36) and (39), Chapter 9. Relation (41) can now be used to index, in terms of p, q, any Laue spot on the cylindrical film that has $| x_c | < \frac{1}{2}\pi r_F$. But for the Laue spots having $| x_c | > \frac{1}{2}\pi r_F$, the diffracted ray intercepts the flat film in the back-reflection region. In this case, (43) of Chapter 9 must be used. For these values of x_c, the sign of x_b is opposite to the sign of x_c. The y_b, on the other hand, satisfies (39) and (44) of Chapter 9.

Other systems. If the crystal is not cubic, it is still possible to make use of the equations (40) and (41) given for the cubic case, provided that indices of the directions $[u_0v_0w_0]$, $[u_1v_1w_1]$ and the planes (hkl) of the noncubic crystal are first referred to a Cartesian orthogonal system. Let us consider the most general case, that is, the indexing of a Laue photograph of a triclinic crystal. Let us choose a Cartesian orthogonal system $OXYZ$, Fig. 8, such that OZ coincides with the triclinic c direction, OY is perpendicular to OZ in the plane containing c and b, and OX is perpendicular to both OY and OZ. If $[uvw]$ is a given direction of the crystal with respect to the triclinic axes, then the direction cosines $[UVW]$ of this direction with respect to the orthogonal system can be calculated in terms of the triclinic constants a, b, c, α, β, and γ^* by

$$U = a \sin \beta \sin \gamma^* u,$$

$$V = -\sin \beta \cos \gamma^* u + b \sin \alpha v, \tag{45}$$

$$W = a \cos \beta u + b \cos \alpha v + cw.$$

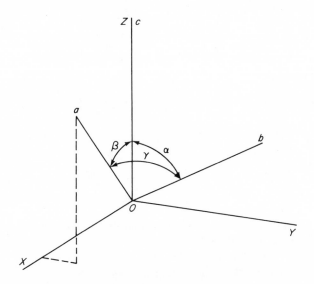

Fig. 8. The triclinic crystal axes referred to a Cartesian coordinate system.

We can write (45) in matrix notation

$$\mathbf{U} = \mathbf{L}u \tag{46}$$

where $\mathbf{L}$ is the lower triangular matrix

$$\mathbf{L} = \begin{vmatrix} a \sin \beta \sin \gamma^* & 0 & 0 \\ -a \sin \beta \cos \gamma^* & b \sin \alpha & 0 \\ a \cos \beta & b \cos \alpha & c \end{vmatrix}. \tag{47}$$

Equations (45) allow one to compute the direction cosines $[U_0 V_0 W_0]$ of the incident x-ray beam with respect to the orthogonal system if $[u_0 v_0 w_0]$ is known. Similarly, one can easily obtain $[U_1 V_1 W_1]$ and $[U_2 V_2 W_2]$ in terms of $[u_1 v_1 w_1]$ and $[u_2 v_2 w_2]$, respectively. These values have to be computed only once for a given Laue photograph.

The next step is to calculate the indices (HKL) referred to the orthogonal axes corresponding to a given (hkl) referred to the triclinic crystal. It can be shown that the crystallographic indices (hkl) of the triclinic crystal will transform into the (HKL) indices referred to the orthogonal system in terms of

$$H = (\mathbf{L}')^{-1}h. \tag{48}$$

The elements of this transformation matrix are equal to the elements of the inverse, transposed matrix **L**. We have

$$(\mathbf{L}')^{-1} = \begin{vmatrix} \dfrac{1}{a \sin \beta \sin \gamma^*} & \dfrac{\cos \gamma^*}{b \sin \alpha \sin \gamma^*} & \left(\dfrac{\cos \alpha \cos \gamma^*}{c \sin \alpha \sin \gamma^*} - \dfrac{\cos \beta}{c \sin \beta \sin \gamma^*} \right) \\[3ex] 0 & \dfrac{1}{b \sin \alpha} & -\dfrac{\cos \alpha}{c \sin \alpha} \\[3ex] 0 & 0 & \dfrac{1}{c} \end{vmatrix},$$

$$(49)$$

which is now an upper triangular matrix. Note that H is parallel to a^* and that K lies in the a^*b^* plane.

The indices (HKL) in the orthogonal system corresponding to a given (hkl) of the triclinic crystal are then given by

$$H = \frac{1}{a \sin \beta \sin \gamma^*} h + \frac{\cos \gamma^*}{b \sin \alpha \sin \gamma^*} k - \left(\frac{\cos \alpha \cos \gamma^*}{c \sin \alpha \sin \gamma^*} - \frac{\cos \beta}{c \sin \beta \sin \gamma^*} \right) l,$$

$$K = 0 + \frac{1}{b \sin \alpha} k - \frac{\cos \alpha}{c \sin \alpha} l,$$

$$L = 0 + 0 + \frac{1}{c} l.$$

$$(50)$$

The elements of the matrix $(\mathbf{L}')^{-1}$ have to be computed just once for a given crystal. Then for every spot of film coordinates $L(x, y)$, the corresponding P, Q can be obtained from (40) and (41), in which the $u_2'v_2'w_2'$ have been substituted by the corresponding $U_2'V_2'W_2'$. Once the index ratios P, Q have been obtained for a Laue spot, then the prime indices HKL of the spot can easily be determined. The true indices hkl of the Laue spot can finally be computed by using (50).

Significant literature

1. E. Schiebold. Beitrage zur Auswertung der Laue-Diagramme. *Naturwiss.* **10** (1922) 399–411.

2. Richard M. Bozorth. The orientation of crystals in electrodeposited metals. *Phys. Rev.* 26 (1925) 390–400.

3. E. Schiebold and G. Sachs. Graphische Bestimmung der Gitterorientierung von Kristallen mit Hilfe des Laueverfahrens. Gesetzmässiges Wachstum von Aluminium-kristallen bei der Rekristallisation. *Z. Kristallogr.* 63 (1926) 34–48.

4. M. Majima and S. Togino. The radiograph of a crystal having a face-centered cubic lattice. *Inst. Physical and Chemical Res. Sci. Papers [Rikogaka Kenkyajo]* 7 (1927) 75–76.

5. M. Majima and S. Togino. The radiograph of a crystal having a body-centered cubic lattice. *Inst. Physical and Chemical Res. Sci. Papers [Rikogaku Kenkyajo]* 7 (1927) 259–260.

6. E. Schiebold. Die Lauemethode. [Vol. I of "Methoden der Kristallstrukturbestimmung mit Röntgenstrahlen"] (Akademische Verlagsgesellschaft, Leipzig, 1932) 173 pages.

7. W. May. Calculation of the stereographic pole figure of the cubic lattice for any given direction. *Proc. Koninklijke Nederlandische Acad. van Wetenschappen* 50 (1947) 548–553.

8. Kathleen Lonsdale. Indexing of Laue photographs of cubic crystals. *Acta Cryst.* 1 (1948) 225.

9. C. G. Dunn and W. W. Martin. The rapid determination of orientation of cubic crystals. *Trans. AIME* 185 (1949) 417–427.

10. R. B. Russell. Crystallographic angles for orthorhombic (alpha) uranium. *Trans. AIME* 197 (1953) 1190.

11. A. Taylor and Sam Leber. Crystallographic angles for hexagonal metals. *J. Metals* 6 (1954) 190–192.

12. W. Vickers. Further contributions to the crystallographic angles for bismuth and antimony. *J. Metals* 9 (1957) 827–828.

13. R. J. Peavler and J. L. Lenusky. Angles between planes in cubic crystals. AIME Inst. Metals Special Report No. 8 (1959) 28 pages.

14. J. D. H. Donnay and Gabrielle Donnay. Cubic interplanar angles. [In John S. Kasper and Kathleen Lonsdale (editors), "International tables for x-ray crystallography," Vol. 2. Kynoch Press, Birmingham, England, 1959] 119–122.

15. R. E. Frounfelker and W. M. Hirthe. Crystallographic data for the tetragonal crystal system. *Trans. AIME* 224 (1962) 196–198.

16. C. G. Wilson. Standard stereographic projection of gallium. *Trans. AIME* 224 (1962) 1293–1294.

17. Elizabeth A. Wood. Crystal orientation manual. (Columbia Univ. Press, New York, 1963) 75 pages.

18. M. Saleh Ahmed and S. Georgeoura. Indexing of asymmetrical Laue photographs of all crystal systems. *Z. Kristallogr.* 118 (1963) 273–282.

19. Y. A. Konnan. A systematic method of indexing spots of single crystals in Laue x-ray photographs. [In "Advances in x-ray analysis," Vol. 7 (Plenum Press, New York, 1964)] 107–116.

20. J. H. Palm. The computation of stereographic projections for non-cubic crystals. *Z. Kristallogr.* 123 (1966) 388–390.

Chapter 12

The characteristic component
of Laue photographs and its application
to the study of diffuse scattering

The intensity distribution of the x-ray spectrum plays an important role in Laue photographs. Even though the Laue method requires a polychromatic x-ray beam, we must recall that the characteristic radiation of the target is by far the most intense in the spectrum. This component has important consequences in Laue photographs. It is the aim of this chapter to recognize and analyze those diffraction phenomena associated with the Laue technique that are produced by the characteristic component of the x-ray spectrum.

Bragg spots in Laue photographs

From a technical point of view, the most distinctive feature of the Laue method is that the crystal is kept motionless during the entire experiment. Therefore, the reciprocal lattice has a fixed orientation with respect to the incident beam, so that diffraction due to characteristic radiation can occur only if some points of the reciprocal lattice lie on the reflecting sphere of radius $1/\lambda_{K\alpha}$. For an arbitrary orientation of the crystal, diffraction would be a fortuitous coincidence. Nevertheless, this kind of diffraction frequently occurs in Laue photographs, producing a few strong, easily recognizable spots. These are the Bragg spots of the Laue photograph. Two features help in the recognition of Bragg spots: first, the Bragg spots have high intensity compared with the average Laue spots; second, the Bragg spots in Laue photographs of oriented crystals appear along layer lines, similar to those observed in oscillation photographs.

The number of Bragg spots in a Laue photograph depends on the probability that a reciprocal-lattice point lies on the Ewald sphere. This probability, in turn, depends only on the density of lattice points in reciprocal space. This density is proportional to $1/v^*$ where v^* is the volume of the primitive cell of the reciprocal lattice. Crystals with small translations, such as simple inorganic compounds, have small densities of lattice points in reciprocal space; accordingly, Bragg spots have a small probability of occurrence. On the other hand, organic compounds, with large unit cells, have high densities of points in reciprocal space, so Bragg spots are frequent in their Laue photographs. An extreme example can be seen in the Laue photographs of stearic acid, where the Laue photograph is almost comparable to an oscillation photograph.

Background of Laue photographs

Bragg spots are not the only consequence of the effect of the characteristic radiation. Real crystals possess disorder of many kinds. This causes diffuse regions of appreciable intensity that are not concentrated only at the nodes of the reciprocal lattice. In such circumstances it is quite likely that the Ewald sphere may intersect one or more such scattering regions, and that the consequent scattering will be recorded on the Laue photographs. The presence of this diffuse scattering is easily recognized in Laue photographs: the diffuse-scattering spots may be associated with strong Bragg spots or they may cover wide regions of the Laue photograph in a cloudlike manner. The study of diffuse scattering is of fundamental importance in understanding a crystal's dynamics[4,7] or its translational or statistical disorder.[5]

In order to understand diffuse scattering of any sort, it must be recognized that the diffraction space of a crystal is not discontinuous, but rather continuous with a modulated intensity that depends on the reciprocal space vector $\mathbf{r}^*$. The Laue photograph is a record of a spherical cut of diffraction space. Therefore, the modulation of the background observed in such photographs is directly related to the modulation of the intensity function in diffraction space.

The x-ray diffuse scattering arising from thermal or statistical disorder in the crystal structure is much weaker than the intensity of the Laue or

Bragg reflections. The diffraction of any wavelength of the polychromatic radiation used in the Laue method gives rise to some diffuse scattering. But the main consequence of the sum of the diffuse scattering due to the wavelengths of the continuous spectrum is to increase the general background of the photograph. The characteristic radiation has high intensity and, consequently, the diffuse scattering that is observed in Laue photographs, superimposed on the general background of uniform intensity, is due only to the contribution of this characteristic radiation. This allows one to study diffuse scattering with the aid of Laue photographs.

Since the x-ray diffuse scattering is produced by the characteristic radiation, it is sufficient to filter the x-ray beam to obtain a substantially monochromatic background due to the $K\alpha$ radiation. Therefore, the interpretation of the diffuse scattering appearing in a Laue photograph is fairly simple: the diffuse scattering is simply the projection on the film of the diffraction space cut by the $K\alpha$ Ewald sphere.

Because diffuse scattering is very weak, the exposure time of the Laue photograph must be long enough to allow it to be recorded. It is obvious that if the crystal remains stationary, as in the Laue method, the exposure time necessary to register diffuse scattering will be much less than if moving-film methods are used for the same purpose. The method that takes advantage of this situation has been called the *monochromatic Laue method*, although a better name is the *characteristic Laue method*. It has been shown[6] that a full record of diffraction space can be made by taking Laue photographs of a crystal in a systematic manner, a method that is called the *systematic Laue method*; this procedure is described in another section of this chapter.

The nature of diffuse scattering as deduced from the study of Laue photographs

The exact nature of diffuse scattering can be deduced only through a painstaking study of the many factors involved. For this purpose the reader is referred to more specialized books.[4,7] However, a qualitative picture of the main factors involved in the nature of diffuse scattering can be easily understood just by comparing appropriately selected Laue photographs. Since this monograph deals with the Laue method only, it is appropriate to show here in pictorial form the basic factors involved.

Monochromatic origin. The first basic question about the diffuse scattering observed in a Laue photograph concerns its monochromatic

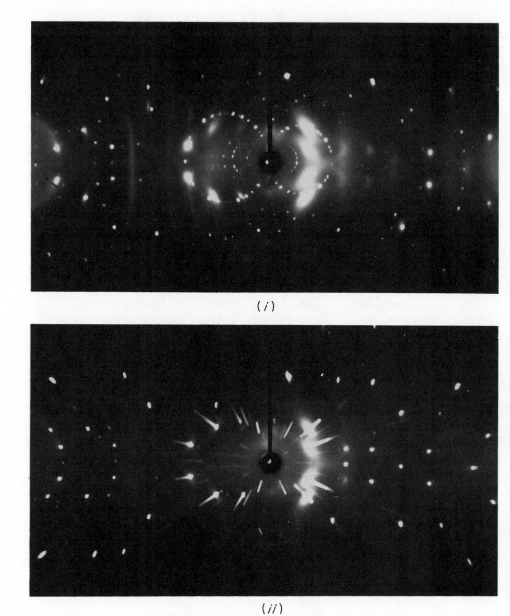

(*i*)

(*ii*)

Fig. 1. X-ray photographs taken with a cylindrical camera of a succinic acid crystal with [010] parallel to the cylinder axis. (*i*) Laue photograph. (*ii*) Oscillation photograph.

nature. A qualitative proof can be obtained by comparing a Laue photograph with a photograph taken by oscillating the crystal a few degrees. Figure 1*i* reproduces a cylindrical Laue photograph of succinic acid, a monoclinic crystal, taken with the x-ray beam at 20° from [100]. Diffuse-scattering regions are clearly shown. A photograph taken in the same orientation but with an oscillation of $7\frac{1}{2}°$ about the cylinder axis is reproduced in Fig. 1*ii*. The Laue spots have disappeared and only Bragg reflections occur. In both photographs, however, the diffuse scattering appears in the same positions. Since the diffraction recorded in the oscillation photograph is due only to the characteristic radiation (except the radial streaks), it follows that the diffuse scattering observed in the Laue photograph is also due only to the characteristic radiation and, therefore, is monochromatic in nature.

A more powerful proof of the monochromatic origin of diffuse scattering is given by comparing a Laue photograph with a photograph taken of the same crystal in the same orientation but using an x-ray beam rendered monochromatic by crystal reflection. This technique was used by Lonsdale and Smith[2]. The diffuse scattering observed in the Laue photograph was identical to that recorded by using a strictly monochromatic x-ray beam.

Diffuse scattering due to temperature. Thermal diffuse scattering depends on the thermal vibrations of the atoms in the crystal. Since the thermal vibrations increase with temperature, this kind of scattering also increases with increasing temperature. As an example, two photographs of hexamine, a cubic crystal, are shown in Figs. 2*i* and 2*ii*. The first one was taken at room temperature, the second one at liquid-nitrogen temperature. It is clear that the diffuse scattering observed at room temperature is much more intense than that at low temperature. The reverse effect is observed with the Laue spots. This shows that the diffuse scattering observed in hexamine is temperature-dependent.

Diffuse scattering due to disorder. A disordering of the structure units of the crystal causes the appearance of diffuse scattering. Its appearance can be shown in the transition from ferroelectric to paraelectric phases. For instance, Fig. 3*i* shows a Laue photograph of $NaNO_2$, an orthorhombic ferroelectric crystal at room temperature. Thermal diffuse scattering can be seen in the Laue photograph. By raising the temperature, we cause the crystal to undergo a phase transition into a paraelectric phase in which the NO_2 groups are statistically oriented along the *c* axis. A Laue photograph taken above the transition point (Fig. 3*ii*) shows diffuse streaks, in addition to the normal thermal-diffuse scattering. These streaks are due to the disorder of the NO_2 groups in the structure.

Dependence of thermal diffuse scattering on structure type. The first aim of crystal-structure determination is to ascertain the type of structure. Some basic features about structure type can be deduced in an easy way from the diffuse scattering, especially from the continuous diffuse scattering in molecular crystals.[7] Such regions are weak and spread over large areas between reciprocal-lattice points, even through lattice points that are forbidden by the space group. We shall summarize here the findings for different types of structure.

1. *Spherical molecules joined by van der Waals forces.* Hexamine is an example. Continuous diffuse regions of spheroidal shape are observed.
2. *Planar molecules joined by van der Waals forces.* Examples: naphthalene, anthracene, and acridine III. In reciprocal space there exist limited, intense regions of continuous diffuse scattering. The anisotropy and finite dimensions of the molecules are revealed in the anisotropic and well-defined finite regions of continuous diffuse scattering. The two orientations of the molecules in the unit cell are shown in the two symmetric orientations of the corresponding diffuse regions.
3. *Elongated molecules joined in chains by hydrogen bonds.* Examples: dicarboxylic acids. In these acids, well-defined continuous diffuse sheets normal to the molecular chains are observed. The shape of these regions of scattering is related to the transform of the molecule.
4. *Molecules joined in layers by hydrogen bonding.* Example: pentaerythritol. The observed diffuse scattering consists of continuous bands normal to the plane of the molecules. The structure within the plane is manifested in the spacing of the bands of scattering, since their spacing is related to the intramolecular interatomic distances.
5. *Molecules joined in a three-dimensional framework.* Example: oxalic acid dihydrate. The molecules have two symmetric orientations, and form clear spiral chains within the structure. The continuous diffuse regions observed are sheets normal to the two directions of the chains and with two different orientations related to the orientations of the molecules in the cell, but there also appear wider regions of diffuse scattering, and the whole phenomenon is of an intermediate type.
6. *Clathrate compounds.* Example: urea-suberic acid. In this case independent diffuse scattering is observed corresponding to each of the two constituents of the structure, namely, molecules that form boxes and molecules that are trapped within them. Thus the reciprocal spacing of

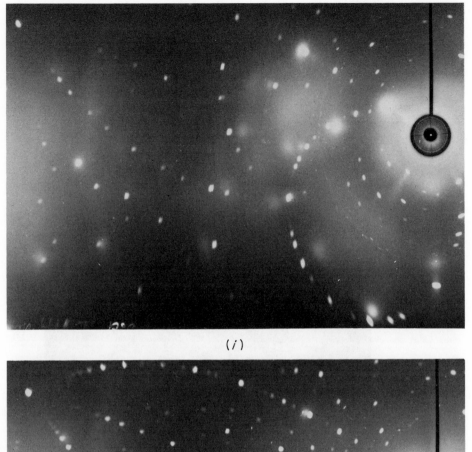

(*i*)

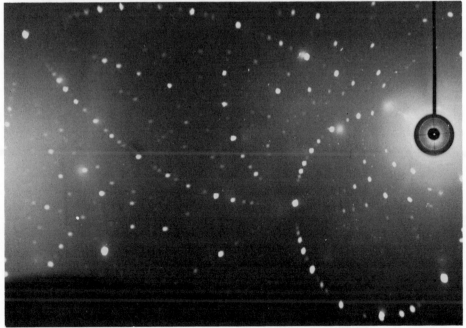

(*ii*)

Fig. 2. Hexamine. Laue photograph at (*i*) 20°C; (*ii*) −160°C.

299

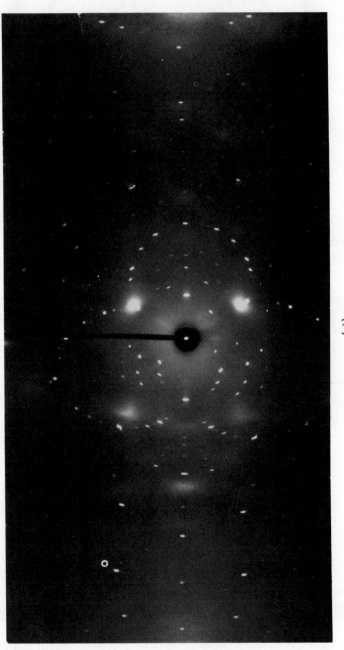

(i)

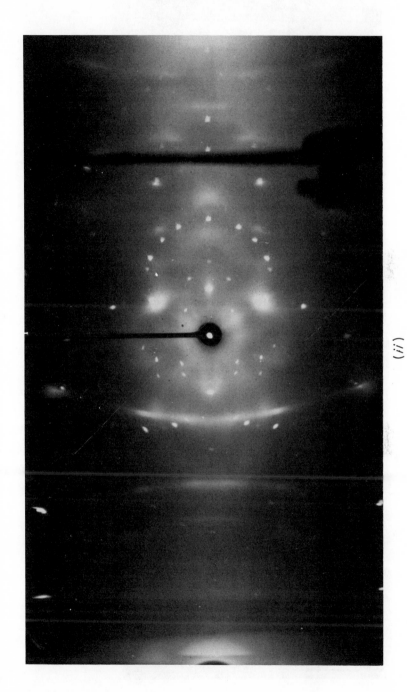

(ii)

Fig. 3. NaNO$_2$. Laue photographs taken with a cylindrical camera. (i) Ferroelectric, ordered structure at 20°C. (ii) Paraelectric, disordered structure at 180°C.

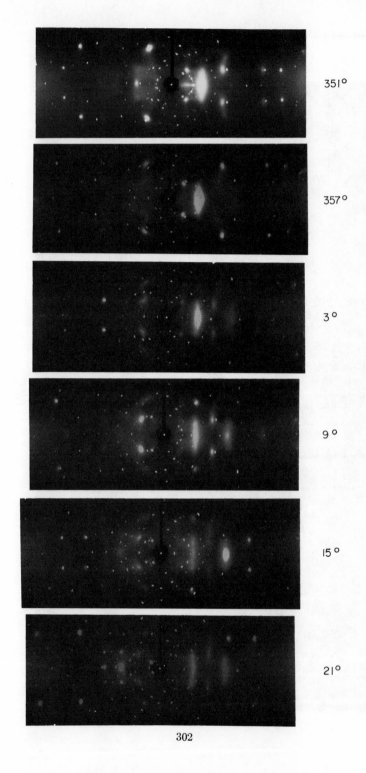

351°

357°

3°

9°

15°

21°

302

the continuous diffuse sheets that correspond to the trapped molecules is independent of the reciprocal spacing of the cage.

Extended diffuse scattering. The relations discussed above imply that the scattering is extended over a distinct volume in diffraction space. This extension is a characteristic of diffuse scattering as compared with Bragg or Laue spots, which have pointlike characteristics. The character of extension of diffuse scattering is dramatically shown in Fig. 4, which consists of a series of Laue photographs of NH_4NO_3 at room temperature taken at successive orientations of the crystal. The orientation of the crystal was changed for each photograph, that is, the crystal was rotated 6° about the vertical axis [100] from one photograph to the next. The angles referred to in the figure are based upon assigning 0° as the orientation of the crystal that fulfills the Bragg condition for 020. It can be observed that the Laue spot (010) migrates from the left to the right of the Bragg reflection 020, indicating the variation of the orientation of the crystal in the different Laue photographs. The diffuse scattering associated to 020, on the contrary, does not change its position but does change its intensity and shape, the highest intensity coinciding with the setting at 0°. The diffuse scattering associated with 020 is so extended in depth that it can even be observed at orientations of the crystal ±30° from the position for the reflection 020. The whole diffuse domain has the shape of a biconvex lens centered on the 020 reciprocal-lattice point with maximum intensity at the lattice point. In a similar way, other domains of diffuse scattering can be observed in this crystal.

The character of extension of diffuse scattering allows one to map it as a continuous function, as will be shown in a later section of this chapter.

Interpretation of the characteristic component of the Laue photograph

Cylindrical coordinates. A cylindrical coordinate system is the most suitable for the interpretation of the characteristic radiation in cylindrical Laue photographs, as it is also for rotation and oscillation photographs.[†] If the axis along which the crystal is mounted is taken as the axis of the cylinder, the cylindrical coordinates of any point of reciprocal space are ξ,

---

[†] M. J. Buerger, X-ray crystallography. (Wiley, New York, 1942) 133–155.

Fig. 4. NH_3NO_4 IV. A series of six cylindrical Laue photographs taken at room temperature, showing the Laue spots and diffuse scattering for varied crystal orientation.

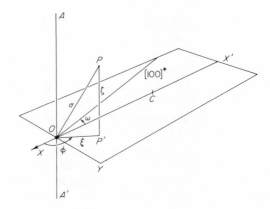

Fig. 5. Representation of the dimensionless reciprocal-space coordinates ϕ, ξ, ζ.

ζ, and ϕ (Fig. 5). Since P' is the projection of P on the plane XOY, the point P can be defined in terms of $\zeta = \overline{P'P}$, $\xi = \overline{OP'}$, and $\phi = XOP'$.

The magnitude of the position vector $\overline{OP}$ can be expressed in terms of the ζ, ξ coordinates

$$\sigma = (\zeta^2 + \xi^2)^{1/2}. \tag{1}$$

This position vector is related to the reciprocal vector $\mathbf{r}^*$ by the wavelength of the x radiation used

$$\mathbf{\sigma} = \mathbf{r}^*\lambda. \tag{2}$$

Since

$$|\mathbf{r}^*| = \frac{2 \sin \theta}{\lambda}, \tag{3}$$

it follows that the magnitude of $\mathbf{\sigma}$ is dimensionless and merely given by

$$\sigma = 2 \sin \theta. \tag{4}$$

In the Laue method the reciprocal lattice is fixed and diffraction due to the characteristic radiation occurs only for points of reciprocal space lying on the sphere of reflection. Points on the sphere can be specified by only the pair of coordinates ζ and ξ. However, in order to define the position in reciprocal space of diffraction points, it is necessary to fix the sphere of reflection in a particular orientation in the reciprocal lattice. This can be specified in terms of an orientation angular coordinate ω, the angle that the x-ray beam forms with an arbitrary axis of the reciprocal lattice (say a^*). This implies fixing the orientation of the reciprocal lattice with respect to the x-ray beam and the reflecting sphere, in which case points lying on the sphere are unequivocally determined in terms of only a coordinate pair ξ,

ζ if care is taken about the signs. The coordinate ω is the angular coordinate common to all diffuse spots appearing in the same Laue photograph, and can be determined in terms of the Laue spots, as we saw in Chapter 9.

Curves of constant ζ and constant ξ on the sphere of reflection. Let us consider the sphere of reflection of radius unity. In Fig. 6, O is the origin of the reciprocal lattice and $\overline{CO}$ the direction of the x-ray beam. The direction of a reflected ray $\overline{CP}$ can be given in terms of the cylindrical coordinates ζ, ξ of the point P on the sphere. In Fig. 6, $\xi = \overline{OM}$ and $\zeta = \overline{MP}$; arc $OA = \Upsilon$ (upsilon) is the azimuth angle, which is normal to the plane containing P and the axis of the camera. Arc $AP = \chi$ (chi) is the latitude of P from this plane, and arc $OP = 2\theta$ is the diffraction angle. We have

$$\zeta = \sin \chi, \tag{5}$$

and, by the law of cosines applied to the oblique triangle CMO,

$$\xi = [2 - \zeta^2 - 2(1 - \zeta^2 \cos \Upsilon)^{1/2}]^{1/2}. \tag{6}$$

Reciprocal-space points having the same value of the ζ coordinate lie on a plane normal to the camera axis; the locus of the contacts of these points with the sphere lies on the intersection of this plane with the sphere of reflection. This intersection is a small circle parallel to the equator of the sphere, of radius given by

$$R = (1 - \zeta^2)^{1/2}. \tag{7}$$

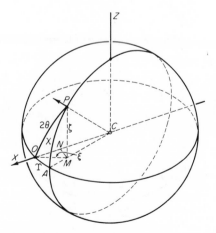

Fig. 6. Spherical coordinates 2θ, χ, and Υ.

Points in reciprocal space having the same value of the ξ coordinate lie on the surface of a cylinder whose radius is ξ. This cylinder intersects the sphere in a closed ovoid curve which represents the locus of points where reciprocal-space points of the same coordinate ξ strike the sphere to produce reflections.

For this arrangement of the x-ray beam and coordinate system, the coordinate ζ is limited to the range -1 to 1. The coordinate ξ is limited between 0 and 2 for the equator (i.e., when $\zeta = 0$) and $0 < \xi < 2$ for $|\zeta| > 0$.

For each value of ζ there is a lower limit of ξ below which diffraction cannot occur, for from (6) it is seen that

$$\Upsilon = \cos^{-1}\left[\frac{2 - \zeta^2 - \xi^2}{2(1 - \zeta^2)^{1/2}}\right]; \tag{8}$$

thus, given a value of ζ, unless ξ has a certain minimum value, the numerator of the right member of the equation exceeds the denominator and $\cos \Upsilon > 1$. When this occurs, $\cos \Upsilon$ becomes meaningless and the reflection has no existence.

The sphere of reflection can be mapped with curves of constant ζ and constant ξ (Fig. 7). It must be pointed out that on the sphere there are four

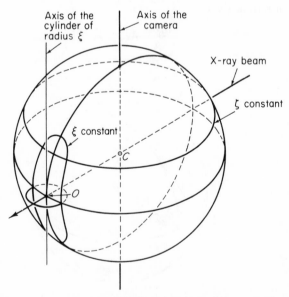

Fig. 7. Mapping of the curves of constant ζ and ξ on the sphere of reflection.

points corresponding with the same absolute value of the ζ, ξ pair. In order to have biunicity between points in reciprocal space and points on the sphere, we must assign signs to the coordinates ζ and ξ, or label the four quadrants of the sphere or upper right $(+\xi, +\zeta)$, upper left $(-\xi, +\zeta)$, lower right $(+\xi, -\zeta)$, and lower left $(-\xi, -\zeta)$. If this criterion is added, there is biunicity between points in reciprocal space satisfying this diffraction condition and reflecting points on the sphere.

The angular coordinate 2θ of a reflecting point on the sphere can also be expressed in terms of the reciprocal coordinates ζ, ξ of this point. From the isosceles triangle OCP (Fig. 6) it follows that

$$\overline{OP} = 2\sin\theta = (\zeta^2 + \xi^2)^{1/2}, \tag{9}$$

so

$$\sin\theta = \frac{(\zeta^2 + \xi^2)^{1/2}}{2}. \tag{10}$$

The sine of the deviation angle 2θ can be obtained from $\sin\theta$ by applying the well-known trigonometric relation

$$\sin 2\theta = 2\sin\theta\cos\theta \tag{11}$$

$$= \tfrac{1}{2}[(\zeta^2 + \xi^2)(4 - \zeta^2 - \xi^2)]^{1/2}. \tag{12}$$

The following relation can also be deduced.

$$\cos 2\theta = \tfrac{1}{2}(2 - \zeta^2 - \xi^2). \tag{13}$$

The relations between reciprocal-space coordinates and film coordinates

The reflections are actually recorded on cylindrical or flat films, so that the pattern of spots on the film is a projection of the reflecting points on the sphere. Every diffuse region on the sphere is projected on the Laue photograph as a diffuse area. The ζ, ξ coordinate pair of any point in the diffuse area on the sphere corresponds with a unique point having coordinates x, y on the film.

Cylindrical Laue photographs. Let us consider a cylindrical film with the cylinder axis normal to the x-ray beam, and the reflecting sphere inside, as in Fig. 8. From the proportionality of arc and angle

$$\frac{x}{2\pi r_\mathrm{F}} = \frac{\Upsilon}{360°}, \tag{14}$$

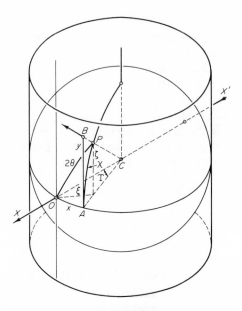

Fig. 8. Geometry of the cylindrical Laue photograph. P, point of the characteristic component on the sphere of reflection; ξ, ζ, dimensionless reciprocal-space coordinates of P; B: x, y coordinates of the characteristic spot on the film.

the film coordinate x is evidently

$$x = \frac{2\pi r_F}{360°}\, \Upsilon. \tag{15}$$

The value of Υ in terms of the reciprocal-space coordinates is given by (8). Substituting this value in (15) gives

$$x = \frac{2\pi r_F}{360°} \cos^{-1}\left[\frac{2 - \zeta^2 - \xi^2}{2(1 - \zeta^2)^{1/2}}\right]. \tag{16}$$

The film coordinate y of the diffuse spot on the Laue photograph is related to the coordinate ζ through the similar-triangle relation

$$\frac{y}{r_F} = \frac{\zeta}{(1 - \zeta^2)^{1/2}}, \tag{17}$$

from which

$$y = r_F \frac{\zeta}{(1 - \zeta^2)^{1/2}}. \tag{18}$$

Combining (6) and (15), we get the value of the reciprocal-space coordinate ξ in terms of ζ and of the x coordinates of the diffuse spot on the Laue photograph

$$\xi = \left[2 - \zeta^2 - 2(1 - \zeta^2)^{1/2} \cos\left(\frac{360°}{2\pi r_F} x\right) \right]^{1/2}. \tag{19}$$

From Fig. 8 we have

$$\sin \chi = \frac{y}{(r_F{}^2 + y^2)^{1/2}}. \tag{20}$$

Combining (20) and (5), we have the reciprocal-space coordinate in terms of the y coordinate of the diffuse spot on the Laue photograph

$$\zeta = \frac{y}{(r_F{}^2 + y^2)^{1/2}}. \tag{21}$$

Flat transmission Laue photographs. Referring to Fig. 9, let us assume that the film is tangent to the reflecting sphere and let D be the crystal-to-film distance. The projection of the portion OP of a great circle

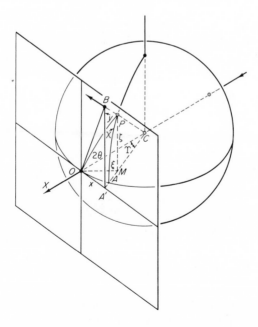

Fig. 9. Geometry of flat Laue transmission photographs.

on the film is OB. OA' is the projection of the portion OA of the equator. We have

$$\tan \Upsilon = \frac{x}{D}. \tag{22}$$

Substituting the value of Υ given by (8) in (20), one gets

$$x = D \tan \cos^{-1}\left[\frac{2 - \zeta^2 - \xi^2}{2(1 - \zeta^2)^{1/2}}\right]. \tag{23}$$

From the similar triangles $BA'C$ and PMC one can write the proportion

$$\frac{A'B}{MP} = \frac{CB}{CP}. \tag{24}$$

If the radius of the sphere of reflection is D, then

$$MP = D\zeta \qquad \text{and} \qquad CO = CP = D.$$

In the rectangular triangle OBC,

$$CB = \frac{CO}{\cos 2\theta} = \frac{D}{\cos 2\theta}.$$

Substituting the appropriate values in (24), there is obtained

$$\frac{y}{D\zeta} = \frac{D/\cos 2\theta}{D},$$

which leads to

$$\frac{\zeta}{\cos 2\theta} = \frac{y}{D}, \tag{25a}$$

from which, finally,

$$y = D \frac{\zeta}{\cos 2\theta}. \tag{25b}$$

Combining (25) and (13), one gets

$$y = D \frac{2\zeta}{2 - \zeta^2 - \xi^2}. \tag{26}$$

Relations (23) and (26) allow us to calculate the x, y coordinates on the film from the reciprocal coordinates ζ, ξ of a diffuse spot. The cylindrical coordinates in reciprocal space can, in turn, be calculated from the film coordinates. From Fig. 9 and by using (5), we have

$$\zeta = \frac{y}{(x^2 + y^2 + D^2)^{1/2}}. \tag{27}$$

Also, from Fig. 9,

$$\cos \Upsilon = \frac{D}{(x^2 + D^2)^{1/2}}, \tag{28}$$

and substituting this value into (6), one gets

$$\xi = \left[2 - \zeta^2 - 2(1 - \zeta^2)^{1/2} \frac{D}{(x^2 + D^2)^{1/2}} \right]^{1/2}. \tag{29}$$

Back-reflection Laue photographs. In this case, a point having coordinates x, y on the film has coordinates ζ, ξ on the reflecting sphere. The relationship between the y coordinate of the film and the reciprocal coordinate ζ is the same as in the transmission case; therefore

$$\zeta = \frac{y}{(x^2 + y^2 + D^2)^{1/2}}. \tag{30}$$

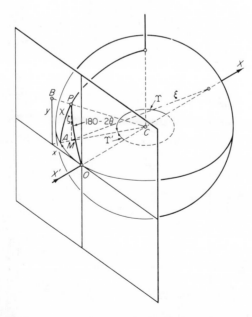

Fig. 10. Geometry of flat Laue back-reflection photographs.

Table 1

Collected transformations between film coordinates and reciprocal-space coordinates

	Cylindrical	Flat transmission	Back reflection
$\xi =$	$\left[2 - \zeta^2 - 2(1-\zeta^2)^{1/2} \cos\left(\dfrac{360°}{2\pi r_F}\right) \right]^{1/2}$ (19)	$\left[2 - \zeta^2 - 2(1-\zeta^2)^{1/2} \dfrac{D}{(x^2+D^2)^{1/2}} \right]^{1/2}$ (29)	$\left[2 - \zeta^2 + 2(1+\zeta^2)^{1/2} \dfrac{D}{(x^2+D^2)^{1/2}} \right]^{1/2}$ (31)
$\zeta =$	$\dfrac{y}{(r_F^2 + y^2)^{1/2}}$ (21)	$\dfrac{y}{(x^2+y^2+D^2)^{1/2}}$ (27)	$\dfrac{y}{(x^2+y^2+D^2)^{1/2}}$ (30)
$x =$	$\dfrac{2\pi r_F}{360°} \cos^{-1}\left[\dfrac{2 - \zeta^2 - \xi^2}{2\,(1-\zeta^2)^{1/2}}\right]$ (16)	$D \tan \cos^{-1}\left[\dfrac{2 - \zeta^2 - \xi^2}{2\,(1-\zeta^2)^{1/2}}\right]$ (23)	
$y =$	$r_F \dfrac{\zeta}{(1-\zeta^2)^{1/2}}$ (18)	$D \dfrac{2\zeta}{2 - \zeta^2 - \xi^2}$ (26)	

In Fig. 10, the angle Υ is related to x as in (8), but Υ' is related to Υ by

$$\Upsilon' = 180° - \Upsilon;$$

therefore

$$\xi = \left[2 - \zeta^2 + 2(1 + \zeta^2)^{1/2} \, \frac{D}{(x^2 + D^2)^{1/2}} \right]^{1/2}. \tag{31}$$

The several transformations for deriving reciprocal-space coordinates ζ, ξ and film coordinates x, y from one another are assembled in Table 1.

The relations that exist between ζ, ξ and x, y for spots that are due to characteristic radiation in Laue photographs are the same as for the rotation and oscillation photographs. Table 1, for instance, is the one given by Buerger[†] as the collected transformations between film coordinates and reciprocal-lattice coordinates, valid for the rotation and oscillation method. In the Laue method, however, instead of restricting the use to reciprocal-lattice points, the equations are used mainly for areas in reciprocal space that are not necessarily confined to reciprocal-lattice points.

The Bernal charts

The most convenient way of interpreting diffuse spots due to the characteristic radiation is to prepare a chart of the film showing the ζ and ξ coordinates of every point directly. Now that the transformations from reciprocal-space coordinates to film coordinates have been developed, the means are available for the preparation of such charts. Given a pair of values ζ and ξ for a reciprocal-space point, the position x, y of the corresponding diffuse spot, as recorded on the film and due to this point, is given by formulas (16) and (21) for a cylindrical film, and formulas (23) and (26) for a flat film.

In order to map out the area of the film in terms of reciprocal-space coordinates, it is only necessary to assume decimal values for the pair ζ, ξ, such as .1, .1;   .1, .2;   .1, .3;   etc., and plot the x, y film coordinates with the aid of (16), (18) or (23), (26) where such pairs of values would record. Connecting all points for which $\zeta = 0.1$, for example, gives a line for which ζ has the constant value 0.1, etc.; and connecting all points for which $\xi = 0.3$, for example, gives a line for which ξ has the constant value 0.3, etc. Proceeding in this way, the coordinates for any spot may be read for the entire area of the film.

Such charts have been prepared by Bernal[1] for both cylindrical and flat

[†] M. J. Buerger, X-ray crystallography. (Wiley, New York, 1942) 144.

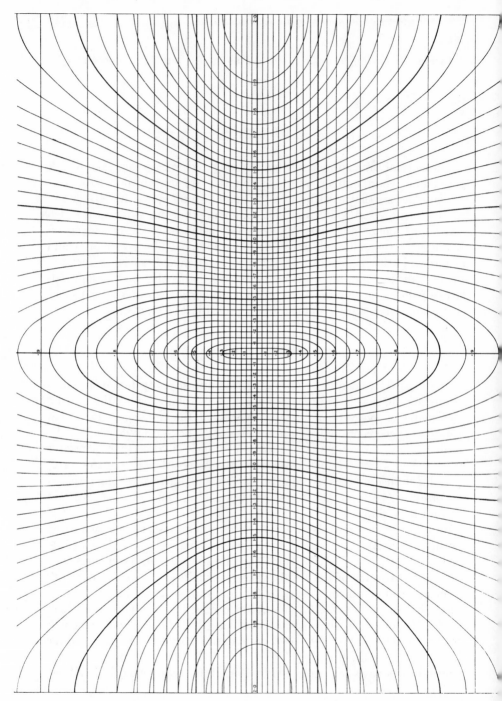

314

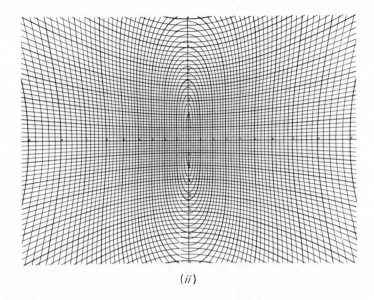

(*ii*)

Fig. 11ii. Bernal chart for reading coordinates of diffuse spots on flat transmission Laue photographs. Crystal-to-film distance, 3.0 mm. [From Bernal[1], Chart I.]

transmission films. These are shown in Fig. 11*i* reduced for a cylindrical camera of 57.296-mm diameter (1 mm per 2° Υ, which is a particularly convenient dimension for all cylindrical-camera work) and in Fig. 11*ii* for a flat transmission plate distant $D = 3$ cm from the crystal (also a particularly convenient dimension). Bernal did not calculate the chart corresponding to the back-reflection case. In practice, it is most convenient to have the chart photographed on a transparent film or plate. The x-ray film is placed over this film and viewed against an illuminated opal-glass background. The reciprocal-space coordinates ζ, ξ may then be read for each diffuse spot with the greatest ease.

The systematic Laue method

A survey of all diffraction space can be obtained from a series of Laue photographs that are made in such a way that the orientation of the crystal is systematically changed. This is, in essence, the systematic Laue method. In this method the crystal under investigation is mounted for rotation

Fig. 11*i*. Bernal chart for cylindrical film for reading coordinates of diffuse spots on Laue photographs. Camera radius, 28.65 mm. [From Bernal[1], Chart II.]

about a known crystal direction, and a Laue photograph is then made with sufficient exposure time to bring out the detail in the diffuse scattering. The crystal is then rotated through a known angle ω to a new orientation, in which another Laue photograph of the same exposure time is made. The procedure is repeated as many times as is necessary to complete the survey of all diffraction space. Flat-film Laue photographs record a smaller angular range than cylindrical Laue photographs, so for this reason the systematic Laue method assumes that cylindrical-film recording is being used. The number of photographs necessary to survey all diffraction space may be considerably reduced by making use of the symmetry of the crystal under study. The precise number of photographs needed to survey the minimum angular range that covers the unique part of diffraction space depends on the degree of resolution required. A convenient increment of ω is $3°$.

The systematic Laue method was used by Amoros and Canut[6] in a semi-quantitative way. It has been shown by McMullan *et al.*[8] that the method is capable of yielding intensity data as accurate as those obtained by the Weissenberg method of recording diffuse-scattering data. Moreover, the systematic Laue method possesses certain advantages in cases where the crystal sublimes during the experiment. For quantitative studies, the method requires correlating the intensity data from one photograph with data obtained from other photographs, a problem that is discussed in a later section of this chapter.

Mapping diffraction space. Once the systematic set of Laue photographs has been obtained, the next step is to reconstruct all of diffraction space by mapping the information contained in the photographs. The most convenient way to deal with this is to map diffraction space in sections at given heights about the equator. This procedure may be undertaken for each orientation of the crystal. This amounts to mapping the information at values of constant ζ; thus, the problem is reduced to a two-dimensional one, in which the plane coordinates are ξ and ω. For a given Laue photograph, ω = constant. The mapping can be done as a function of ξ by plotting the observed diffuse scattering at different levels, each with constant ζ, and then rotating the Ewald sphere systematically by $\Delta\omega$ in the direction opposite to that in which the crystal was rotated to take the next Laue photograph.

The mapping of diffraction space is greatly simplified by using a chart in which the curves of constant ξ and constant ζ on the Ewald sphere are projected in orthogonal form on its equator. A chart of this kind, reproduced in Fig. 12, was presented by Martin[3]. The curves of constant ξ are pro-

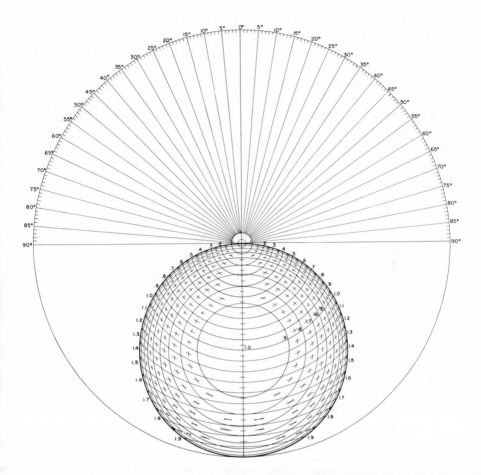

Fig. 12. Orthogonal projection of the sphere of reflection mapped with the constant-ξ and constant-ζ curves. [From Amorós and Canut de Amorós[6], p. 307, Fig. 192; after Martin[3], p. 515, Fig. 1.]

jected as a set of circles concentric with the center of the Ewald sphere, and the curves of constant ζ are another set of circles concentric with the origin of diffraction space. The Martin chart is drawn normally with 10 cm as the radius of the Ewald sphere. The set of concentric circles, each corresponding to constant ζ, are given for values of $\zeta = 0.1, 0.2, \ldots, 0.9$; identical increments are used for the circles of constant ξ. The distance from the center of the chart to any point in the projection is $2 \sin \theta$. A circle of 20-cm radius is also drawn with its center at the origin of diffraction space; a scale of degrees drawn in this circle serves to provide the angle ω.

The Martin chart is drawn on bond paper; superimposed on it is a sheet of transparent paper to be used to map the observed diffuse scattering for a given ξ in the Laue photograph along the corresponding ξ circle in the chart. By rotating the transparent paper, the information from each Laue photograph of the systematic set can be plotted, and thus the whole diffraction space can be mapped. For the set of six Laue photographs of NH_4NO_3 IV reproduced in Fig. 4 the camera axis was [100] and the orientation interval 6°. A portion of the kind of map obtained from the photographs in Fig. 4 is given in Figs. 13 and 14. The circles in Fig. 13 represent the successive cuts

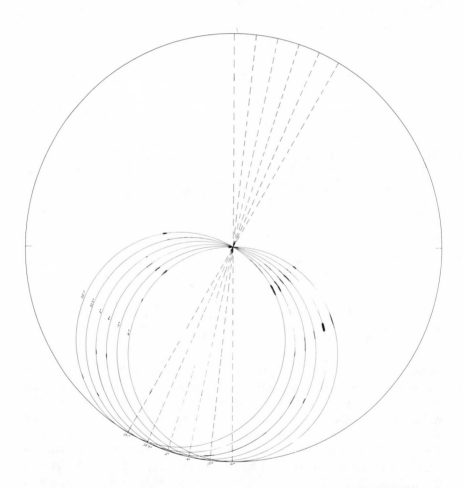

Fig. 13. Schematic representation of the equator of the sphere of reflection at different crystal orientations. These intersections correspond to the Laue photographs of Fig. 4.

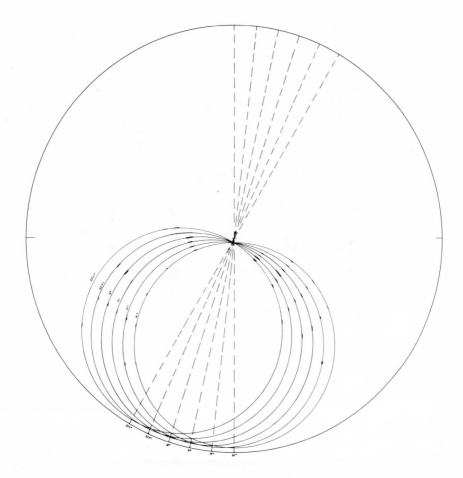

Fig. 14. Schematic representation of the section of the sphere of reflection on level $\zeta = 0.17$ at different crystal orientations. These correspond to the Laue photographs of Fig. 4.

of the equator of the Ewald sphere (i.e., at $\zeta = 0$) by the successive Laue photographs. The small circles in Fig. 14 represent the successive cuts of the Ewald sphere at $\zeta = 0.17$. Due to the symmetry of the crystal, only a limited series of Laue photographs is normally needed, especially if the information from $+\xi$ and $-\xi$ is used as given in each Laue photograph. The complete maps of the equator and the first level parallel to $(100)^*$ are shown in Fig. 15. The actual set of Laue photographs used for these mappings consists of 30 Laue photographs taken at intervals of $3°$.

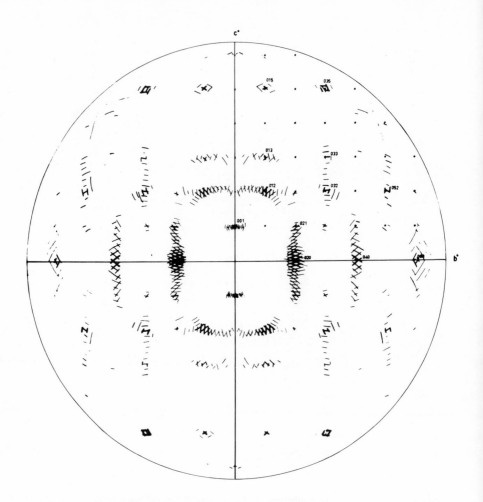

Fig. 15. Diffuse scattering of NH_3NO_4 IV as obtained by the systematic Laue method: the zero level for [100] parallel to the axis of the camera.

Blind zones. The volume of reciprocal space surveyed by the systematic Laue method is determined, first, by the wavelength used; this determines the radius of the sphere of reflection. The second factor involved is the size of the photographic film.

Let us consider the case of the cylindrical Laue technique. If we have the crystal so oriented that [010] is parallel to the camera axis, the whole plane (010)* is recorded up to the limiting sphere (Fig. 16*i*). The upper levels, however, are limited by the size of the film; for a Unicam camera

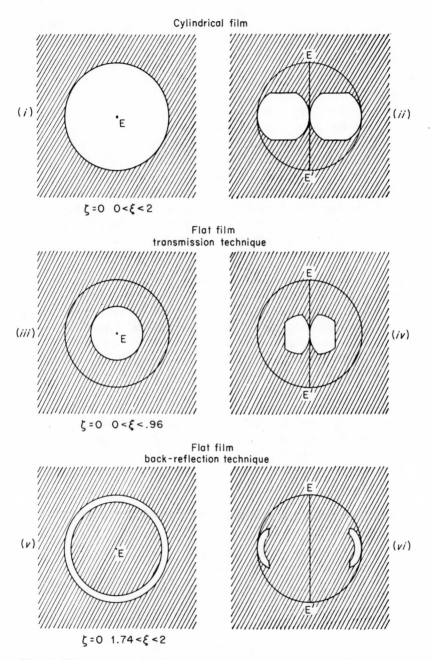

Cylindrical film

(i) $\zeta = 0$ $0 < \xi < 2$

(ii)

Flat film
transmission technique

(iii) $\zeta = 0$ $0 < \xi < .96$

(iv)

Flat film
back-reflection technique

(v) $\zeta = 0$ $1.74 < \xi < 2$

(vi)

Fig. 16. Blind zones of the characteristic component as registered by the different Laue techniques. *EE'*; axis of the camera. [From Amorós and Canut de Amorós[6], p. 309, Fig. 194.]

(length 120 mm) levels with $\zeta > 0.9$ are not recorded. Figure 16*ii* represents the region recorded for any section perpendicular to (010)* and containing the origin of reciprocal space.

If, instead of cylindrical film, the flat transmission technique is used, the limitation on the diffraction space surveyed is severe. For instance, for the Unicam flat film the record is limited to a distance of 48 mm from the record of the direct beam. This value corresponds to a maximum value of $\xi = 0.96$; beyond this no information is available. The region surveyed in (010)* is that given in Fig. 16*iii* and the transverse section is given in Fig. 16*iv*.

In a similar way, Figs. 16*v* and 16*vi* represent the surveyed regions for the back-reflection technique. It is evident that the cylindrical Laue technique allows the optimum survey of reciprocal space.

Interfilm calibration of intensities. The greatest advantage of the systematic Laue method is that, since the entire section of diffraction space defined by the position of the Ewald sphere and the crystal orientation is recorded simultaneously, no layer-to-layer correlation problem exists. A problem does exist, however, with the correlation between different films of the set. The solution to this problem has been provided by the following simple method, developed by McMullan and co-workers.[8]

Consider two Laue photographs taken at some angular interval to be designated $2\omega'$. Because of the reciprocity of motions of the sphere of reflection and the crystal, we can imagine that the sphere has been moved from position 1 to position 2. In fact, the crystal was rotated $-2\omega'$ for the second photograph. In Fig. 17 the equatorial sections of the sphere in these two positions are shown. The equation of the sphere, of unit radius, centered at C is

$$\xi^2 + \zeta^2 - 2\xi \cos (\omega - \omega_1) = 0, \tag{32}$$

and that of the plane of intersection of the two spheres is

$$\omega = \tfrac{1}{2}(\omega_2 - \omega_1) = 0. \tag{33}$$

The intersection of the two spheres centered at 1 and 2 is a circle given by

$$\xi^2 + \zeta^2 - 2\xi \cos \omega' = 0, \tag{34}$$

since

$$\omega' = \tfrac{1}{2}(\omega_2 - \omega_1). \tag{35}$$

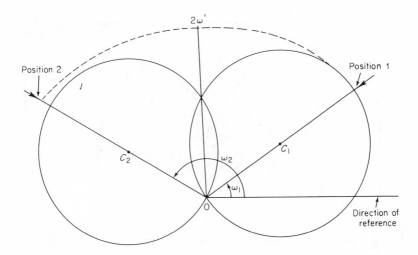

Fig. 17. The intersection of the characteristic spheres of reflection for two Laue photographs taken at an angular separation of $2\omega'$.

The circle (34) contains all the diffracting points common to both Laue photographs. The diffracted beams passing through the perimeter of this circle fall on an oval curve on each Laue photograph, and, evidently, the corresponding points on the two ovals have the same intensity, if both Laue photographs have been obtained in the same conditions. The parametric equations

$$x = R \cos^{-1} \left\{ \frac{1 - \xi \cos \omega'}{(1 + \xi^2 - 2\xi \cos \omega')^{1/2}} \right\}, \tag{36}$$

$$y = R \left\{ \frac{2\xi \cos \omega' - \xi^2}{1 + \xi^2 - 2\xi \cos \omega'} \right\}^{1/2} \tag{37}$$

of the points on the oval curves can be readily derived from (15) and (18). These oval curves are the zone ellipses that were described in Chapter 8; here they are used to correlate the intensities of any two Laue photographs. The location of the oval curves on the film is a function of ω', a situation that is illustrated in Fig. 18, where the curves for $\omega' = 12.8°$, $37.0°$, $60.0°$, and $75.5°$ are represented. In order to correlate the intensities of the films it is only necessary to compare the points on the same oval curves on both films. Because the procedure is independent of the sequence of films, and

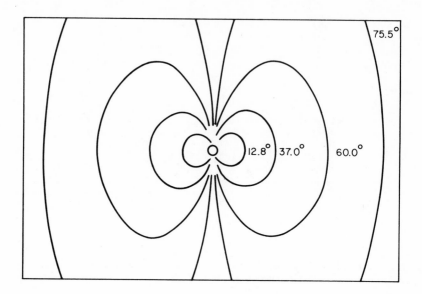

Fig. 18. Intersection of the zone cones in a cylindrical Laue photograph. The curves correspond to $\omega' = 12.8°$, $37.0°$, $60.0°$, and $75.5°$, respectively, starting from the center. [From McMullan *et al.*[8], p. 295, Fig. 5.]

depends only on the angular distance $2\omega'$ of the two films, the most appropriate oval curves can be selected by comparing pairs of Laue photographs not necessarily contiguous in the systematic Laue set.

The use of symmetry in mapping diffraction space. The power of the systematic Laue method is greatly enhanced when full use is made of the symmetry of diffraction space. It not only provides more points for the interfilm calibration, but also greatly reduces the angular range that is necessary for recording the information of all diffraction space. However, not all the elements of symmetry that may exist in diffraction space of a given crystal can be used. In fact, only those that contain the camera axis or that are perpendicular to it are suitable. In other words, due to Friedel's law, the centrosymmetry of all diffraction space can be used, as well as the presence of axes of symmetry parallel to or perpendicular to the camera axis, and the mirror planes either perpendicular or parallel to the camera axis. It is left to the reader as an exercise to elaborate this problem, which is schematically indicated in Table 2.

Table 2

Utilization of symmetry in the systematic Laue method [after McMullan et al.,[8] modified]

Crystal system	Friedel symmetry	Crystal setting: crystal axis which is parallel to cylinder axis	Diagram showing symmetry-generated spheres of reflection	Minimum range of ω	Restrictions on initial direction of x-ray beam
Triclinic	$\bar{1}$	[uvw]		180°	None
Monoclinic	$\dfrac{2}{m}$	About 2-fold axis (c axis for first setting, b axis for second setting.)	As for $\bar{1}$, but no vertical polarity	180°	None
		About a or c axis (second setting)		90°	Along b, a, or c

Table 2 (continued)

Crystal system	Friedel symmetry	Crystal setting: crystal axis which is parallel to cylinder axis	Diagram showing symmetry-generated spheres of reflection	Minimum range of ω	Restrictions on initial direction of x-ray beam
Orthorhombic	$\dfrac{2}{m}\dfrac{2}{m}\dfrac{2}{m}$	About a, b, or c axis	As for $2/m$ about a or c axis, but no vertical polarity	90°	Along a, b, or c
	$\dfrac{4}{m}$	About c axis		90°	None
		About a_1 axis	As for $2/m$ about a or c	90°	Along c or a_2
Tetragonal	$\dfrac{4}{m}\dfrac{2}{m}\dfrac{2}{m}$	About c axis		45°	Along a_1, a_2, or c
		About a axis	As for $2/m$ about a or c axis but no vertical polarity	90°	Along c or a_2

System	Point group	Axis	Pattern	Angle	Polarity
Hexagonal	$\bar{3}$	About 3-fold axis		60°	None
		Any other axis	As for $\bar{1}$	180°	None
	$\bar{3}\dfrac{2}{m}$	About 3-fold axis		30°	Along 2-fold axis, or parallel to mirror plane
	$\dfrac{6}{m}$	About 6-fold axis	As for $\bar{3}$, but no vertical polarity	60°	None
	$\dfrac{6}{m}\dfrac{2}{m}\dfrac{2}{m}$	About 6-fold axis	As for $\bar{3}\,2/m$, but no vertical polarity	30°	Along 2-fold axis, or parallel to mirror plane
Cubic	$\dfrac{2}{m}\bar{3}$	About a_1 axis	As for $2/m$ about a or c axis, but no vertical polarity	90°	Along a_2, a; or [011]
		About [111]	As for 3	60°	None
	$\dfrac{4}{m}\bar{3}\dfrac{2}{m}$		As for $\dfrac{4}{m}\dfrac{2}{m}\dfrac{2}{m}$ about c axis	45°	Along a_2, a; or [011]
		About [111]	As for $3m$	30°	Along 2-fold axis, or parallel to mirror plane

Significant literature

1. J. D. Bernal. On the interpretation of x-ray, single crystal, rotation photographs. *Proc. Roy. Soc. (London)* **A113** (1926) 117–160.
2. K. Lonsdale and H. Smith. An experimental study of diffuse x-ray reflexion by single crystals. *Proc. Roy. Soc. (London)* **A179** (1942) 8–50.
3. W. G. Martin. Improved method for plotting reciprocal lattice points. *J. Appl. Phys.* **27** (1956) 514–515.
4. W. A. Wooster. Diffuse x-ray reflections from crystals. (Clarendon Press, Oxford, 1962) 200 pages.
5. A. Guinier. X-ray diffraction in crystals, imperfect crystals, and amorphous bodies. (Freeman, San Francisco, 1963) 378 pages.
6. J. L. Amorós and M. L. Canut de Amorós. La difracción difusa de los cristales moleculares. (Consejo Superior de Investigaciones Cientificas, Madrid, 1965), 351 pages, especially pp. 299–311.
7. José Luis and Marisa Amorós. Molecular crystals: their transforms and diffuse scattering. (Wiley, New York, 1968), 479 pages.
8. J. T. McMullan, D. M. Burns and W. G. Ferrier. Correlation of intensity measurements of the diffuse scattering of x-rays. *J. Appl. Cryst.* **1** (1968) 293–298.

Chapter 13

The optics of the Laue method

In the preceding chapters we have been dealing with the interpretation of the Laue diagram as if it were a photograph made with a point crystal. However, crystals have finite sizes, and x-ray beams are far from being composed of parallel rays, even in the best-collimated cases. As a result of these experimental features, the shape, size, and intensity distribution of the spots in a Laue photograph are affected. A study of these effects involves the optics of the Laue method.

The shapes of Laue spots

The spots in a Laue photograph are just the images of the crystal as transmitted through the optical system defined by the experimental setup. The spots vary in size and form, depending on a variety of factors. First, there are the intrinsic causes derived from the experimental method itself, namely, the effects due to divergence of the incident beam and the distribution of intensity in it, as well as to the shape of the film used in taking the photograph. Other effects are related to the specimen, that is, to the shape and size of the crystal, its absorption, its extinction, and its distortion or imperfection; the last two characteristics involve a difference in behavior between the surface layers and the bulk of the crystal. The crystal may also show distortions, incurred during growth or due to possible subsequent mechanical action, which produce relatively large disorientation of regions in the specimen. The character of the spot depends on the orienta-

329

tion of the imperfection relative to the crystal plane that produces a given Laue reflection. The shape and structure of Laue spots are therefore far from being uniform, so a great deal of information can be obtained by analyzing each Laue spot separately. This information is not straight-forward, but requires a careful analysis of the geometry of the optics of the Laue method.

In analyzing a single Laue spot it must always be kept in mind that the spot has often received contributions from two or more different wave-lengths reflected in different orders. If the beam is not parallel and the crystal is not small, or if the planes in the crystal deviate from strict parallelism, the Laue photograph reveals these features by the deviation of the shape of the Laue spots from what is expected from simple geometri-cal construction. When a Laue photograph is taken of a crystal with anisotropic habit, the shape of the Laue spot is determined by the orienta-tion of the crystal relative to the film and the incident beam (ignoring any nonparallelism of the beam and absorption by the crystal). Simple sketches serve to indicate that if the incident rays are not quite parallel, the actual dimensions of the Laue spot depend on the crystal-to-film distance and the nature of the nonparallelism of the rays. On Laue photographs taken at any distance, a focusing of some Laue spots and a spreading of others occur. This effect was shown and interpreted as early as 1913 by Bragg[1]. Simple drawings also show that the tendency of a divergent beam is to broaden the Laue spots in back-reflection techniques.

In some cases, when the crystal is large enough to cover the whole divergent beam, the Laue spots may not have sharp edges because of the penumbra region of the beam due to the collimator. In other cases, a sharp intense spot of well-defined shape is superimposed upon a larger, fainter spot, not necessarily concentric with the bright spot. Lonsdale[5] showed that this situation occurs when a Bragg spot is generated for that particular orientation of a given crystallographic plane.

Absorption may cause double spots on the Laue photograph. This effect is frequently observed in Laue spots from crystals having great absorption and produced by long wavelengths, especially the Bragg reflections. It is, however, more frequently due to the mosaic character of the outer layers of the crystal, as will be shown in the next section.

The shape of the focal spot on the target of the x-ray tube also has a strong influence on the shape of the Laue spots when the beam is not fully collimated.

The basic components of a Laue spot

In the Laue method, a well-collimated x-ray beam strikes the crystal. In x-ray diffraction, a collimator consists essentially of two parallel dia-

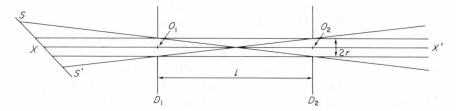

Fig. 1. Optics of a collimator of length l, diaphragm apertures $2r$, and aperture angle 2α.

phragms D_1 and D_2, normally with circular apertures. Figure 1 shows the usual arrangement of a collimating system in which each of the two diaphragms D_1 and D_2 has an aperture of the same diameter, $2r$. The line XX' is the optical axis of the system. For good collimation, $2r$ must be small in comparison with the length $l = O_1O_2$. Such a collimator defines an x-ray beam of nonuniform intensity. Let us assume a large uniform source of x rays SS'. The cylinder limited by the two circular apertures is known as the cylinder of full illumination. There is also the cone of penumbra defined by the half angle α such that $\tan \alpha = 2r/l$. Since α is small, the maximum divergence of the system is $2\alpha \simeq 4r/l$.

If one desires to use the whole aperture of the collimator, including the cone of penumbra, it is necessary that the focal spot of the x-ray tube be large enough to include the whole ellipse defined by the intersection of the cone of penumbra with the plane of the target of the tube. This is seldom accomplished. In the following discussion, we shall assume that the x-ray source is simply bigger than the intersection of the cylinder of full illumination with the plane of the target.

Let us analyze the form of a Laue spot in the case of diffraction of a cylindrical beam. We shall suppose, for the sake of simplicity that, as in Fig. 2, the crystal is in contact with the second diaphragm D_2 of the collimator and more than covers it, and that the film F is flat. We further suppose that the crystal planes are perpendicular to the plane of the drawing and that diffraction occurs throughout the thickness of the crystal. In this case, the radial dimension of the Laue spot is a function of the thickness e of the crystal and the angle θ of diffraction; in Fig. 2

$$\overline{AC} = \overline{AB} + \overline{BC}$$

$$= 2r + e \tan 2\theta. \tag{1}$$

According to this simple geometry, the Laue spot has a complex structure due to the fact that, on the spot produced by the bulk diffraction of the crystal, the effects of the diffraction by the front and back surfaces of the crystal are superimposed. In general, the structure of the surfaces is more imperfect than the structure of the bulk volume of the crystal, and according to the dynamic theory of x-ray diffraction, the intensity of the diffraction from the surface layers of the crystal is higher than that of the bulk volume. Therefore, the Laue spot of a real crystal has higher intensity in the farther and nearer parts of the spot (in relation to the x-ray beam). The components of the outer surfaces of the crystal overlap only when, in (1)

$$\tan 2\theta \leq 2r/e. \tag{2}$$

When this occurs the center of the Laue spot has higher intensity than the outer parts of the spot. In general we can say that the Laue spots are more intense when the surface texture of the crystal is less perfect; however, the definition of the spot is greater when the surface is more perfect.

According to this discussion, the Laue spot has a structure that depends on the overlap of the three components, namely, the reflection due to the surface of entry, the reflection of the bulk of the crystal, and the reflection of the surface of exit. The highest intensity is normally observed in both the first and last components because the mosaic structure is normally more evident in the surface than in the bulk of the crystal. The influence of distortion on the intensity reflected has been used to show the location of imperfections such as dislocations. The Berg-Barrett[6] or the Lang[12, 13] techniques, which use this effect, may be considered as special modifications of the Laue technique.

The structure of the Laue spot depends, therefore, on the ratio of the width of the crystal to the aperture of the diaphragms of the collimator. Different structures of Laue spots are represented in Fig. 3; Fig. 3*i* corresponds to a crystal whose thickness is large in relation to the diameter of the collimator, and Fig. 3*ii* shows an intermediate case. It can be easily seen that, in the experimental conditions sketched in Fig. 2, the position of the farther edge of the Laue spot for a given Laue reflection is independent of the thickness of the crystal. Barraud[10] has further shown that it is also independent of the divergence of the beam; consequently, that author suggests using these properties for measuring the diffraction angle with high accuracy.

The six types of Laue diagrams

The x-ray beam defined by the collimator system is, in fact, a bundle of either convergent or divergent rays that can be treated as a cone-shaped beam. The form and shape of the Laue spots depend on this fact. The different possibilities were examined in detail by Leonhardt[3]. He assumed

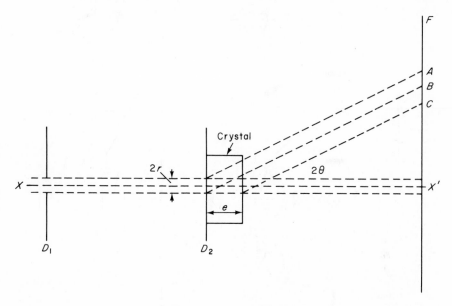

Fig. 2. Geometry of the structure of the Laue spot, as a function of *e* the thickness of the crystal, *r* the radius of the collimator and 2θ the Bragg deviation angle. [From Barraud[10], p. 233, Fig. 3, modified.]

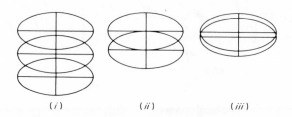

Fig. 3. Structure of the Laue spot as a function of the thickness *e* of the crystal for (*i*) a thick crystal; (*ii*) a crystal of intermediate thickness; (*iii*) a thin crystal. [From Barraud[10], p. 252, Fig. 9, modified.]

that the reflection is produced by an infinitely thin crystal plate, so that the result is dependent only on the conditions of convergency or divergency of the beam and the size of the bundle at the incidence point. Leonhardt distinguishes between large and small surfaces of illumination. The small surface corresponds to a point, in contradistinction to a large area, which has at least one direction of finite dimension. According to Leonhardt the resulting conventional Laue photographs can be classified into the following six types.

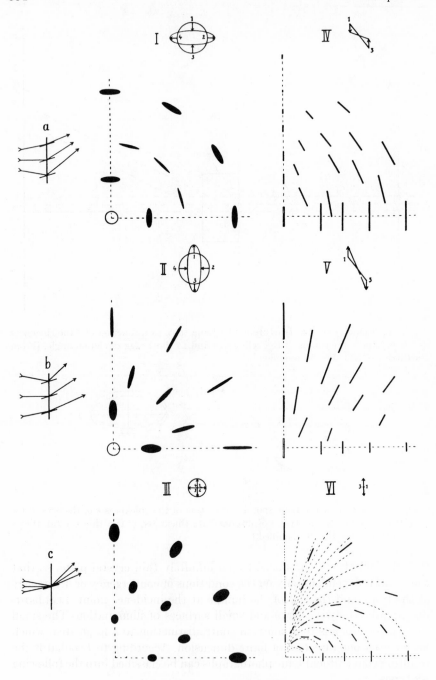

Type I. Conic divergence, large surface of illumination. This is the case treated by Bragg[1]. In the case of a beam of absolutely parallel rays, the diffracted beam has a circular cross section. The effect of the divergence of the beam is to contract the reflected beam along the direction 1–3 (Fig. 4,I) and to expand it along the direction 2–4. The result is that the Laue spots are ellipses with the shortest axis along the radial direction of the spot (the radial direction is the one that connects the spot with the primary beam on the Laue photograph). The largest axis of the ellipse is perpendicular to the radial direction. It can be seen that at a certain angle 2θ, a condition of focusing occurs, so that the Laue spots are almost linear.

Type II. Conic convergence, large surface of illumination. The convergence of the path of the x-ray beams is the reverse of that discussed under type I. The reflected beam expands along direction 1–3 and contracts along direction 2–4 (Fig. 4,II). The ellipses are thus elongated in the radial direction. The excentricity of the ellipses increases with 2θ. No focusing occurs.

Type III. Conic convergence, pointlike surface of illumination. The reflected beam expands in all directions from its center; the intersection of the beam with the film results in an elliptical spot whose size increases with 2θ, (Fig. 4,III).

No other types of conventional Laue photograph can be obtained with a conical beam. However, three more cases are possible when a linear beam is used instead of a conical beam. In that case the following types result.

Type IV. Linear divergence, large surface of illumination. This case was treated also by Gross[2]. A divergent incident beam produces reflections that are contracted along direction 1–3. This is schematically shown in Fig. 4,IV, which also shows that a change in the direction of the line of the reflected beam is produced. This variation depends on 2θ.

Type V. Linear convergence, large surface of illumination. The convergent beam produces reflections that are expanded along direction 1–3, as shown in Fig. 4,V.

Type VI. Linear convergence, pointlike surface of illumination. In this case the points 1 and 3.(Fig. 4,VI) coincide in a single point in the crystal. The reflected beams diverge to produce the result represented in the figure.

Fig. 4. The six types of Laue diagrams. [From Leonhardt[3], p. 481, Fig. 2 and pp. 484–485, Figs. 4–9.]

The optics of the conventional Laue method

The conventional Laue method uses a collimating system with two circular apertures D_1 and D_2 that normally have the same radius r. The line passing through the centers of the two circular apertures is the optical axis of the system. In the optical system so defined, the entrance aperture D_1 of the collimating system plays the role of the object. Let us assume for simplicity that the brilliance of the object is uniform and that the crystal is an infinitely thin layer lying on the aperture D_2. The paths of the rays for a given Laue reflection are shown schematically in Fig. 5. The reflecting plane in the crystal is, in fact, a stack of parallel planes. Consequently, the observed Laue reflection is the result of the individual reflections of the rays coming from the object D_1 as they are reflected by the different planes of the stack at D_2; these can be treated as the formation of the image of the object by a stack of parallel mirrors. As a consequence, each point m in the crystal reflects a ray. The greatest deviation occurs for the ray coming from the point d_1; these are the only ones that need to be considered in studying the formation of the image. If θ is the glancing angle of the stack of plane (hkl), any individual plane of the stack reflects the rays coming from d_1 at an angle given by $2\theta + \beta$ where β is the angle of deviation (from parallelism to the optical axis) from d_1 to the point y_m whose reflection is considered.

Figure 5 indicates that rays issuing from d_1, and diffracted by points M_1

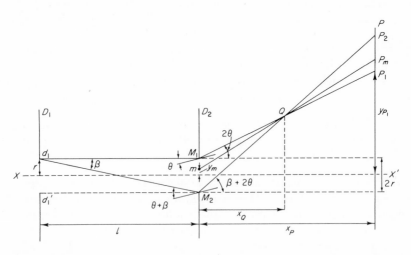

Fig. 5. Paths of the rays in the conventional Laue method. [From Barraud[10], p. 269, Fig. 13, modified.]

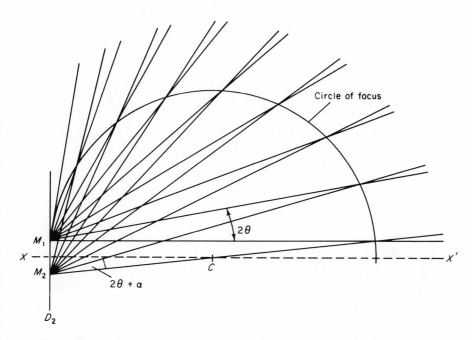

Fig. 6. Geometrical construction illustrating that two nonconcentric pencils of rays, one at intervals of 2θ and the other at intervals of $2\theta+\alpha$, produce a set of intersections on a circle, the circle of focus.

and m of the stack of crystal planes, converge at some point Q. It is an interesting and important feature of Laue photographs that the locus of this convergence, for various values of θ and for various values of α, is a circle in the plane of the diagram, as illustrated in Fig. 6. This circle is called the *circle of focus*. That the locus is a circle follows immediately from the converse of the well-known theorem in geometry that the angle made by two straight lines intersecting on the circumference of a circle is measured by half the arc they intersept. In Fig. 6, rays converging from the upper point M_1 of aperture D_2 and a point at the optical axis are seen to converge along the circumference of a circle that is approximately tangent to the aperture D_2 at its center. Exact focusing occurs from all points of the crystal lying on the arc M_1M_2 of aperture D_2, so that this holds for all α. Since Fig. 5 is a section through a radially symmetrical collimator system, it follows that this focusing occurs in three dimensions, so that focusing actually occurs on the surface of a *sphere of focus*.

The positions of the edges of the diffracted beam as it reaches the plane P

of the film can be found as follows. The angle β is given by

$$\tan \beta = \frac{r \pm y_m}{l} \tag{3}$$

where y_m is the coordinate of the point m measured from the optical axis of the system and l is the separation of the apertures of the collimator.

It is evident from Fig. 5 that the intersection of the reflected rays with the plane P of the photographic plate is given by

$$y_P = x_P \tan(2\theta + \beta) + y_m. \tag{4}$$

By substituting in (4) the value of y_m given in (3), we obtain

$$y_P = x_P \tan(2\theta + \beta) + l \tan \beta + r \tag{5}$$

where l and r are the length and radius of the collimator, respectively. The angle β depends on the position of the point in the object; it has two extreme values defined by the value of the tangent, that is, $\tan \beta = 0$ or $\tan \beta = 2r/l$. For the first case

$$\tan 2\theta = \frac{y - r}{x} \tag{6}$$

and for the second

$$\tan\left(2\theta + \arctan\frac{2r}{l}\right) = \frac{y + r}{x}. \tag{7}$$

The distance x_Q at which the focusing occurs depends on the angle 2θ, and it is given, according to Barraud[10], by

$$x_Q = l \cos^2 2\theta - r \sin 4\theta \tag{8}$$

From Fig. 5 it can be seen that for large enough distances x_P the lowest (proximal) edge of the Laue spot is produced by the ray reflected at M_1, and consequently

$$y_{P_1} = x_P \tan 2\theta + r. \tag{9}$$

Since r is small with respect to y_{P_1}, (9) can be simplified to

$$y_{P_1} = x_P \tan 2\theta \qquad \text{or} \qquad \tan 2\theta = \frac{y_{P_1}}{x_P}, \tag{10}$$

which indicates that the best measurement of 2θ can be obtained by measuring the distance from the proximal edge of the Laue spot to the primary beam. This procedure was recommended by Barraud.[10] For a distance x_P

shorter than x_Q the highest (distal) edge of the Laue spot has to be used. The existence of focusing introduces an aberration in the optical system that is zero at the sphere of focus and increases for observation positions at either side of the point of focus.

Geometry of Laue-spot formation with only one circular aperture

Focusing. In order to deduce the geometry of Laue-spot formation, it is convenient to consider the crystal as infinitely thin. It will be further assumed that the beam is limited by a circular aperture D_2 and that the crystal plate coincides with the plane of the aperture.

Let us consider a point S on the line focus of the target of the x-ray tube, Fig. 7, the line focus being parallel to a line in the plane (hkl) that reflects the x-ray beam. The length $M''M'$ of the plane (hkl) is illuminated by rays coming from the point S and reflected to the points $P''P'$. It can be easily seen that all the x rays coming from the actual source line $S''S'$ and reflected in (hkl) according to the Bragg law seem to come from a virtual source $\sigma''\sigma'$, symmetric to $S''S'$ by reflection across (hkl). It is convenient to use the optical language that S is the object and P is the image. To each beam coming from a point of the object space there corresponds a divergent beam in the image space. It is obvious that for reflection to occur by the Bragg law, each ray of the beam must have a different wavelength because the angle θ is different in each case.

In order to simplify the analysis of the image formation in this case, let us consider the path of the x rays in the plane that contains the optical axis

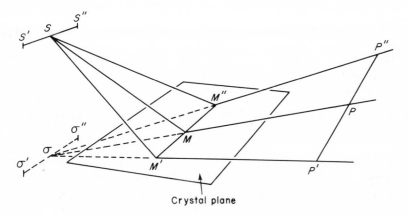

Fig. 7. Geometrical paths of rays reflected by a single plane.

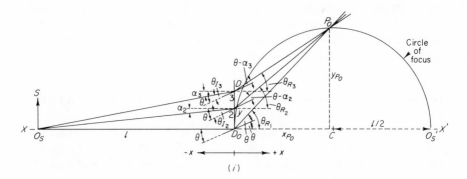

Fig. 8. Formation of the image of a line OS through rays reflected by a plane (hkl). (*i*) Pencil of rays coming from a point O_S on the optical axis.

of the system, Fig. 8*i*; this plane is perpendicular to (hkl). The source of x rays in the plane of the figure is O_sS, and it is at a distance l from the plane of the aperture D_oD. The optical axis of the system is given by the direction O_sD_o. In the figure, three planes of the stack (hkl) are represented and are numbered 1, 2, 3. The trace of this plane forms an angle θ with the optical axis.

The geometry of the image formation can be derived by considering the path of the rays emanating from the two end points of the source, that is, O_s and S, respectively. Every individual plane of the stack (hkl) acts as an individual mirror for which the usual law of reflection holds, namely, that the angle of incidence is equal to the angle of reflection. Let us consider first a pencil of rays whose source is O_s, Fig. 8*i*. A ray O_sD of this pencil forms an angle α_i with the optical axis. The incident glancing angle θ_{Ii} is given by

$$\theta_{Ii} = \theta - \alpha_i \tag{11}$$

and the angle of reflection θ_{Ri}, measured from the direction of the optical axis, is given by

$$\theta_{Ri} = \theta + (\theta - \alpha_i) = 2\theta - \alpha_i. \tag{12}$$

For a stack of planes parallel to the optical axis, $\theta = 0$, and consequently

$$\theta_{Ii} = -\theta_{Ri} = \alpha_i. \tag{13}$$

The pencil of rays coming from O_s converges after reflection into the point O_s' on the optical axis at a distance from D_o equal to l, the distance from the x-ray source to the crystal. For any other angle θ, the pencil of rays reflected by the stack (hkl) converges into a point P_o on the circle of focus whose diameter is l, as in the previous case.

The individual rays coming from the other end S of the source (Fig. $8ii$) make angle α_j with the optical axis. The angle of incidence on the respective planes of the stack is $\alpha_j + \theta$, and the reflected rays form angles $2\theta + \alpha_j$ with the optical axis. Let us make the distance from the target to the aperture, measured along the optical axis,

$$l = -x_s \tag{14}$$

and further, let the abscissa of a point P_s of the image be x_{P_s}. Then the equation of the reflected ray is provided by

$$\tan(2\theta + \alpha_j) = \frac{y - y_j}{x},$$

so that

$$y - y_j = x \tan(2\theta + \alpha_j) \tag{15}$$

where y_j is the y coordinate of the reflecting crystal plane. Since

$$y_S - y_j = l \tan \alpha_j,$$

we have

$$y = y_j + x\tan(2\theta + \alpha_j) - l \tan \alpha_j \tag{16}$$

where α_j is a function of y_S and y_j.

The reflected rays do not converge to a point. This is a well-known situation in optics where stigmatic systems are considered. In this case the pencil of quasi-converging rays can be defined by its evolvent, known in optics as the *caustic*. It is also known that a pencil of rays refracted by a line has a conic for a caustic,[†] a situation to which our example applies.

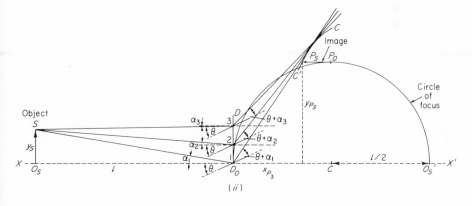

Fig. 8. (ii) Pencil of rays coming from a point S not on the optical axis.

[†] See for instance, Joaquin de Vargas y Aguirre. Catalogo general de curvas, (Memorias Real Acad. Ciencias, Madrid, 1908) 324.

Applying the Herschel condition[14] to the formation of the caustic, there is obtained[10]

$$\frac{x}{\cos^2(2\theta + \alpha_j)} = \frac{l}{\cos^2 \alpha_j},\tag{17}$$

from which the following parametric equations of the caustic can be obtained.

$$x_{P_c} = l\,\frac{\cos^2(2\theta + \alpha_j)}{\cos^2 \alpha_j}.\tag{18}$$

From the triangle mNP_c (Fig. 8*iii*) it follows that

$$NP_c = x_{P_c} \tan(2\theta + \alpha_j).\tag{18a}$$

Taking into account that

$$MN = y_m = y_s - \tan \alpha_j,$$

and substituting in (18a) x_{P_c} given in (18), there is finally obtained

$$y_{P_c} = l\,\frac{\sin(2\theta + \alpha_j)\cos(2\theta + \alpha_j)}{\cos^2 \alpha_j} - l \tan \alpha_j + y_s.\tag{19}$$

By eliminating α_j, one obtains

$$y_{P_c} = \frac{-x_{P_c}\cos 2\theta + 2(Lx_{P_c})^{1/2} - l\cos 2\theta}{\sin 2\theta} + y_s.\tag{20}$$

The caustic is a parabola whose opening is directed toward the axis xx'.

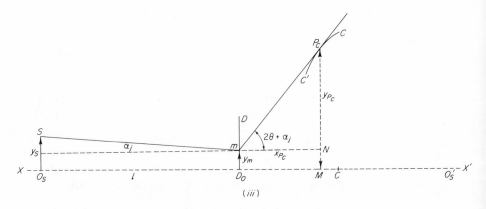

Fig. 8. (*iii*) Coordinates of a point of the caustic.

The image of the source is then $P_o P_S$; it is tangent to the circle of focus in the point of focusing P_o. Let us analyze further the relations existing between the coordinates of the points in the image and the angles θ and α_j. If the aperture tends to zero, y_j also tends to zero, and y_s to $l \tan \alpha_j$. In this case (18) and (19) transform into

$$x_{P_s} = l \, \frac{\cos^2(2\theta + \alpha_j)}{\cos^2 \alpha_j} \,, \tag{21}$$

$$y_{P_s} = l \, \frac{\sin(2\theta + \alpha_j) \cos(2\theta + \alpha_j)}{\cos^2 \alpha_j} \,. \tag{22}$$

If y_s tends to zero, x also tends to zero, and the coordinates are given by

$$x_{P_s} = l \cos^2 2\theta - y_s \sin 4\theta, \tag{23}$$

$$y_{P_s} = l \sin 2\theta \cos 2\theta + y_s \cos 4\theta. \tag{24}$$

For a point on the optic axis where $y_s = 0$, (23) reduces to

$$x_{P_o} = l \cos^2 2\theta. \tag{25}$$

Taking into account that $\sin 2\alpha = 2 \sin \alpha \cos \alpha$, we see that (24) reduces to

$$y_{P_o} = l \, \frac{\sin 4\theta}{2} \,. \tag{26}$$

The preceding equations show that when θ changes, the image P of S moves[4, 10] along a circle of radius $l/2$, tangent to DD_0 at D_0. This circle is the circle of focus that corresponds in space to a sphere of focus.

Taking into account the divergency of the rays in the direction perpendicular to the sagittal plane represented in Figs. 8i and 8ii, the reflected beam is an astigmatic beam, and consequently the image and the focusing are distorted. The distortion increases with the size of the aperture.

Two practical consequences can be deduced from the existence of a sphere of focus:

1. It is possible to obtain extremely fine Laue spots with the aid of a microfocus x-ray tube using only a circular diaphragm, a method due to Poittevin[7, 8].

2. By using a microfocus x-ray tube, it is possible to separate the Laue spots of two crystallites that differ very little in orientation, merely by allowing a sufficiently large distance of focalization, a method due to Guinier and Tennevin[9].

The image field that has just been derived corresponds to an individual

Laue spot, and must not be confused with the field of the whole diagram. The diagram is formed by Laue spots of different indices *hkl*, and therefore to each spot there corresponds a different optical instrument, formed by a different stack of planes (*hkl*). The Laue spots are focused on the sphere of focus. In the conventional Laue method, the film is a flat surface perpendicular to the optical axis of the camera. The film cuts the image space through a plane and consequently only a circular section of the sphere of focus is registered. It is only in this circle that the focusing of the Laue spots can be observed.

The condition for focusing is also a condition of achromatism because in the focusing beam, to each value $\theta + \Delta\theta$ of the Bragg angle there corresponds a wavelength $\lambda + \Delta\lambda$ where $\Delta\lambda$ is a function of $\Delta\theta$. It is important also to note that an x-ray beam of wavelength $\lambda + \Delta\lambda'$ is reflected by the crystal plane (*hkl*) at an angle $\theta + \Delta\theta'$ only when the pair of values $\lambda + \Delta\lambda'$ and $\theta + \Delta\theta'$ satisfy the Bragg equation.

Enlargement of the image field. To complete the study of the image field in the optical system just described, it is necessary to investigate the enlargement of the image field. For this purpose it is useful to consider, following Barraud[10], a small plane element of the source. The small element is perpendicular to the x axis (the direction of the optical axis) and has its x_s dimension constant with a small dimension y_s perpendicular to the optical axis (Fig. 9). From (23) and (24), the following relations are obtained,

$$dx = -\sin 4\theta \, dy_s, \tag{27}$$

$$dy = +\cos 4\theta \, dy_s. \tag{28}$$

In this case the image of dy_s is a small element di such that

$$di = (dx^2 + dy^2)^{1/2} = dy_s; \tag{29}$$

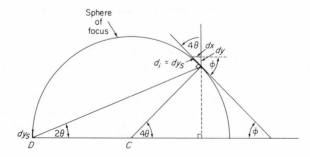

Fig. 9. Components of an element di of the image.

this makes an angle ϕ with the optical axis, which is given by

$$\tan \phi = \frac{dy}{dx}. \tag{30}$$

Therefore, the image di of dy_s is tangent to the sphere of focus. In this sphere the enlargement G is always unity, independent of the focal distance; that is

$$G = \frac{di}{dy_s} = +1 \tag{31}$$

The distribution of intensity in a Laue spot

Just as the shape and size of the Laue spot depend on instrumental and specimen conditions, so does the distribution of intensity in the spot. Not only is the intensity a consequence of the overlap of the components of the reflection, but it is also influenced by instrumental conditions such as divergence of the beam. A detailed investigation due to Barraud[10] will be followed in this section. The experimental assumptions are the ones already discussed in a previous section in this chapter.

Let us define an orthogonal reference system such that the X axis is along the optical axis of the instrument. The focal plane of the source of x rays is perpendicular to the X axis and parallel to the crystal plate and the photographic plate (Fig. 10).

Let us assume that the focal plane of the source of x rays is formed by infinitely thin linear elements parallel to the Z axis. The image of each of these linear elements in the plane of the photographic plate is also a linear element whose length differs from that of the original one due to the astigmatism of the optical system. It has been shown previously in this chapter that if the focal plane of the x-ray source is perpendicular to the optical axis of the instrument, the plane of the image is tangent to the focalization sphere. In this particular situation the lengths of source and image elements are equal. Assuming that all the source elements are identical and have uniform brilliancy, then the illumination at the central line of the spot image (the Laue spot) is constant. However, the intensity decreases from the center to the border of the spot according to a function that depends on the form of the aperture D to which the crystal is attached.

Due to the horizontal divergency of the cone of x rays, an element of dimensions dy_D, dz_D in the crystal gives rise to an element of dimensions

$$dz_P = dz_D(1 + \cos 2\theta) \tag{32}$$

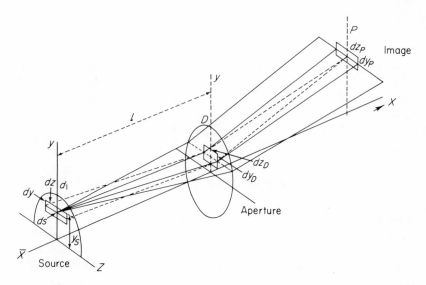

Fig. 10. Paths of rays and elements in the source, aperture, and image. [From Barraud[10], p. 265, Fig. 12, modified.]

at the point of focus. Equation (32) merely takes into account that the dimension of the figure is proportional to the distance from the source. Let us remember that the distance of the crystal from the source is l and that the total path length from the source to the point of focus is $l + l \cos 2\theta$ (Fig. 11).

In accordance with (31), the enlargement of the y dimension at the focal point is unity; then

$$dy_P = dy_D \tag{33}$$

in the image span at the point of focus. The image is formed by a small element dP, such that

$$dP = dz_P \, dy_P = dy_D \, dz_D (1 + \cos 2\theta). \tag{34}$$

The element in the crystal receives the radiation from the x-ray source. We have to consider then the flux of radiation received by this element of the crystal. For this purpose, we consider (Fig. 10) a small element $dS = dy \, dz$ in the source, whose brilliance is B. The flux is given by

$$d^2\phi = B \, dy \, dz \, \frac{dy_D \, dz_D}{l^2} . \tag{35}$$

The illumination of the element of the image is then given by

$$dE = \frac{d^2\phi}{dP} = \frac{B}{l^2} \frac{dy\, dz\, dy_D\, dz_D}{dy\, dz_D(1 + \cos 2\theta)} \tag{36}$$

$$= \frac{B}{l^2} \frac{dz\, dy_D}{(1 + \cos 2\theta)}, \tag{37}$$

assuming that both dz_D and dy_D are small. The total illumination E is given by the integration over all the illuminated surface of the crystal,

$$E = \frac{B}{l^2} \frac{1}{\cos 2\theta} \iint dy_D\, dz_D = \frac{\pi B}{\cos 2\theta} \left(\frac{r}{l}\right)^2 \tag{38}$$

where r is the radius of the circular aperture. This equation is only valid for the image at focus.

The illumination of the Laue spot changes if, instead of using a single aperture, as in the previous case, a collimating system with two identical circular apertures is used. According to Barraud,[10] the illumination is then given by

$$E \sim \frac{4(2)^{1/2}}{(r)^{1/2}} \left(\frac{r}{l}\right)^2 \frac{\cos 2\theta}{1 + \cos 2\theta} B(\Delta u)^{1/2} \tag{39}$$

where l is the length of the collimator, r the radius of the apertures, B the brilliance of the focus, and Δu the distance from the edge of the Laue spot. Conditions for focusing are assumed. The illumination is then represented by the function of Δu that is a parabola with horizontal axis; this means that it is a curve whose tangent is vertical at the origin. As a consequence, the intensity grows rapidly with Δu, meaning that the edges of the Laue spot are very sharp at the plane of focus. This condition can be used for accurate measurement of reflection angles.

All that has been said refers to a Laue spot for which focusing is ac-

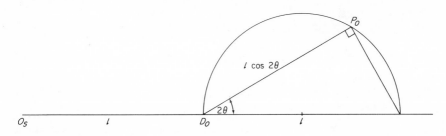

Fig. 11. Path length of the rays from the source to the point of focus.

complished, that is, to all those spots observed at a distance

$$d \sim l \cos^2 2\theta_0 \tag{40}$$

where θ_0 is the angle of the focal center. However, not all Laue spots are recorded at the focus, and consequently the previous equations cannot be applied to all Laue spots without modification. In the case of two apertures, the focusing for Laue spots with $\theta_1 \neq \theta_0$ is accomplished, not for the object defined by aperture D_1 in Fig. 2, but rather by a fictitious object at a distance from D_2 given by

$$l_1 \sim \frac{d}{\cos^2 2\theta_1}. \tag{41}$$

Assuming that the brilliance of this object is the same as that of the source, l_1 is substituted for l in (38), and the illumination at the center of the Laue spot is given by

$$E_1 = \pi \left(\frac{r}{l_1}\right)^2 \frac{B}{\cos 2\theta_1}. \tag{42}$$

By substituting in (41) the value found in (42), we obtain

$$E_1 = \pi B \left(\frac{r}{d}\right)^2 \cos^3 2\theta_1. \tag{43}$$

This can be expressed as a function of E by

$$E_1 = E_0 \left(\frac{\cos 2\theta_1}{\cos 2\theta_0}\right)^3. \tag{44}$$

Equation (44) indicates that, for all other factors constant, the illumination of a Laue spot decreases rapidly with increasing θ. A Laue spot that is focused only for $\theta = \theta_0$ has a higher intensity for $\theta < \theta_0$. An example is given in Table 1. This effect is seen in conventional Laue photographs where the intensities of the spots decrease rapidly with θ.

Table 1

Variation of the intensity of a Laue spot in relation to the intensity at the focus for $\theta_0 = 10°$
[After *Barraud*.[10]]

θ	0°	5°	10°	15°	20°	25°	30°	35°	40°
E_1/E_0	1.19	1.16	1	0.77	0.54	0.32	0.15	0.05	0.006

Significant literature

1. W. L. Bragg. The diffraction of short electromagnetic waves by a crystal. *Phil. Soc. Proc.* **17** (1912) 43–57.
2. R. Gross. Verfestigung und Rekristallisation. *Z. f. Metallkunde* **16** (1924) 344–352.
3. J. Leonhardt. Über den Einfluss von Divergenz und Konvergenz des Primärstrahls auf Form und Grösse der Beugungsflecken im Lauephotogramm. *Z. Kristallogr.* **63** (1926) 478–495.
4. Charles S. Barrett and Carl E. Howe. X-ray reflection from inhomogeneously strained quartz. *Phys. Rev.* **39** (1932) 889–897.
5. Kathleen Lonsdale. The shape of reflections of x-ray, single-crystal photographs. *Mineralog. Mag.* **27** (1945) 112–125.
6. Charles S. Barrett. A new microscopy and its potentialities. *Trans. Amer. Inst. Min. (Metall.) Engrs.* **161** (1945) 15–64.
7. Maurice Poittevin. Tubes à rayons X à foyer linéaire de grande brillance. *Compt. rend. Acad. Sci. (Paris)* **224** (1947) 1709–1711.
8. Maurice Poittevin. Tubes à rayons X à foyer linéaire de grande brillance. *J. de Phys.* **8** (1947) 102–105.
9. A. Guinier and J. Tennevin. Sur deux variantes de la méthode de Laue et leurs applications. *Acta Cryst.* **2** (1949) 133–138.
10. Jean Barraud. Optique de la méthode de diffraction des rayons X de Laue. Application à la mesure précise des paramètres cristallins et à la détermination de l'orientation des cristaux. *Bull. Soc. franç. Minéral. Cristallogr.* **74** (1951) 223–372.
11. L. G. Schulz. Method of using a fine-focus x-ray tube for examining the surface of single crystals. *J. Metals* **6** (1954) 1082–1083.
12. A. R. Lang. A method for the examination of crystal sections using penetrating characteristic x-radiation. *Acta Met.* **5** (1957) 358–364.
13. A. R. Lang. Direct observation of individual dislocations by x-ray diffraction. *J. Appl. Phys.* **29** (1958) 597–598.
14. Max Born and Emil Wolf. Principles of optics. (Macmillan, New York, 1964) 808 pages, especially 168–169.

Chapter 14

Epilogue: Graphical interpretation of the Laue method

The Laue method can be interpreted in an exact way by using a very simple geometrical construction, comparable to the famous Ewald construction. The x-ray spectrum in the Laue method is a continuous set of wavelengths ranging from a well-defined lower limit λ_{min} to an ill-defined maximum. For convenience, let us assume the λ_{max} is ∞; then the whole diffraction space inside a sphere of radius $1/\lambda_{min}$ is filled with a continuous sequence of smaller spheres tangent at the origin of reciprocal space. The outer sphere is the *limiting sphere* of reflection of the Laue method. Any reciprocal-lattice point inside the limiting sphere is in the condition for reflection by selecting the proper smaller sphere of appropriate wavelength. In the present connection, however, we are not interested in the particular wavelength causing the reflection but rather in the angle of reflection.

A reciprocal-lattice row passing through the origin of reciprocal space is perpendicular to the crystal plane that causes the reflection to occur. In the Laue method all the reciprocal-lattice points on a row produce reflections at the same 2θ angle, as noted in Chapter 5, Fig. 13. This means that the direction of reflection is determined by the direction of the reciprocal-lattice row alone. The direction of reflection is given by the line from the center of the limiting sphere to the intersection of the reciprocal-lattice row with this sphere. This is the common direction of the reflections due to every one of the reciprocal-lattice points that form that row and are contained inside the

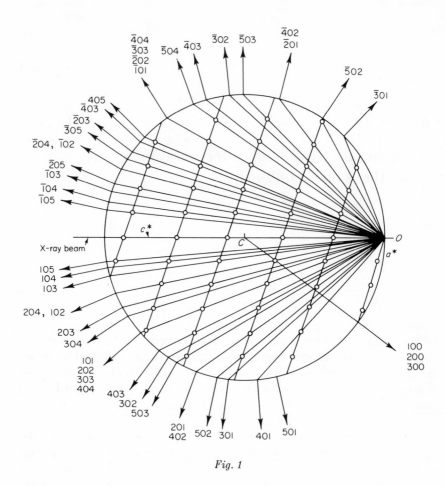

Fig. 1

limiting sphere. A graphical display is given in Fig. 1 where a section of the reciprocal lattice containing a^* and c^* of a monoclinic crystal is represented. All reciprocal-lattice rows that contain the origin O of diffraction space are drawn. The intersections of these rows with the limiting sphere are the windows through which the photons emerge from the point C in the sphere.

Polychromatic and monochromatic Laue reflections

The indices of the emerging reflections are also indicated in Fig. 1. It can be observed that some rays contain a superposition of the rays of several orders of diffraction. These rays are therefore polychromatic. Others, however, are strictly monochromatic, as they are picked up by only one reciprocal-lattice point.

Some interesting features can be deduced from Fig. 1. It can be noticed that the polychromatic Laue rays correspond to reflections from planes with small indices. They usually have high intensity due to the superposition of several orders of diffraction in the same Laue spot. These reflections are surrounded by blank areas, free from any reflection around them, a fact evident in Fig. 1. This has been used to identify them as reflections with small indices by inspection of the Laue photograph. It is evident from the figure that each reflection immediately surrounding the blank area is due to one reciprocal lattice point; consequently, each such reflection is due to diffraction by a specific wavelength alone.

Blank areas

Figure 1 also shows that the Laue spots very near the primary x-ray beam are due to points lying on a reciprocal-lattice plane passing through the origin. This fact can be used in the determination of the translations in this plane provided the orientation of the crystal is not such that this plane is tangent to the limiting sphere. Finally, the figure shows the existence of blank zones like the one between the 100 and 501 reflections; similar regions can be observed near the incident beam in many Laue photographs. These blank areas are due to the discontinuity of reflections from a zero level to a first level of the reciprocal lattice in the direction back along the incident beam.

General equation of the Laue method

The reflection angle $2\theta_i$ of every reciprocal lattice point in the ith reciprocal-lattice row is determined by the inclination angle ϕ_i of that row with respect to the direction of the incident beam. The reflection produced by a lattice point P_1 in this row is due to a particular wavelength λ_1. From Fig. 2 the condition for a reflection by a lattice point P_1 at the end of reciprocal vector $r_1{}^*$ can be derived by using the Ewald construction and substituting θ for its value in (1), as follows.

$$\cos \phi_1 = \frac{r_1{}^*}{2/\lambda_1}. \tag{1}$$

Figure 2 shows that

$$2\theta_i = 180° - 2\phi_i. \tag{2}$$

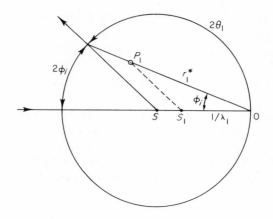

Fig. 2

Since

$$r^*_{nh\ nk\ nl} = 1/d_{nh\ nk\ nl}$$

$$= n/d_{hkl}, \tag{3}$$

(1) can be transformed into

$$\lambda_1 = \frac{1}{r_1^*} 2 \cos \phi_i \tag{4}$$

$$= \frac{2d_1}{n} \cos \phi_i. \tag{5}$$

In general, the relation among λ, d, and ϕ is

$$\lambda = \frac{2d}{n} \cos \phi. \tag{6}$$

This equation is the equivalent of the Bragg equation for the Laue method.

Angular range of a Laue reflection

A reflection by a lattice point P_1 can be observed for a range of orientations ϕ of the lattice row that contains it; $\phi_{\min}$ is obviously zero, and $\phi_{\max}$ occurs when P_1 is on the limiting sphere defined by $\lambda_{\min}$ of the spectrum. From (1) it can be deduced that

$$\cos \phi_{1\ \max} = \frac{\lambda_{\min}}{2} r_1^*, \tag{7}$$

or

$$\phi_{1\,max} = \text{arc cos}\left(\frac{\lambda_{min}r_1{}^*}{2}\right) \tag{8}$$

$$= \text{arc cos}\left(\frac{n\lambda_{min}}{2d_1}\right). \tag{9}$$

The angle ϕ is related to λ through

$$\phi_\lambda = \text{arc cos}\left(\frac{n\lambda}{2d}\right). \tag{10}$$

This equation allows one to determine the setting angle ϕ for selecting the reflection due to a given λ.

Range of wavelengths for a Laue reflection

It is clear from (4) that the wavelength used in causing the Laue reflection at $2\theta_1$ by a reciprocal-lattice point at the end of vector r^*, is a function of ϕ_1. For a particular orientation ϕ_1 of the row, the reflection produced by the particular crystal-lattice plane is strictly monochromatic, involving λ_1. For another angle ϕ_2 the same plane selects another specific wavelength λ_2 to produce a reflection along a new direction $2\theta_2$. The new wavelength is always available because the Laue method uses the continuous spectrum of wavelengths produced by the x-ray source. Not all the available wavelengths are used by the same crystal-lattice plane, however. The spectrum used covers only a range of wavelengths that extends from λ_{min}, determined by the voltage applied to the x-ray tube, to a $\lambda_{1\,max}$ specific to r_1^*. The latter value can be determined by setting $\phi_i = 0$ in (4), in which case

$$\lambda_{i\,max} = \frac{2}{|\,r^*\,|}. \tag{11}$$

The smaller $|\,r^*\,|$ is or the larger $1/d$ is, the wider the range of wavelengths used by the plane.

The Laue method makes use of photographic film. The sensitivity of this photographic film depends on the wavelength of the x rays. There is a sudden increase of sensitivity for wavelengths lower than $\lambda(Br) = 0.918 \times 10^{-8}$ cm and another even sharper increase for wavelengths smaller than $\lambda(Ag) = 0.485 \times 10^{-8}$ cm. The first discontinuity is due to the existence of frequencies $\nu > \nu(Ag)$ that excite the K radiation of silver.

The radiations capable of exciting the K radiation of bromine are those between $\lambda(\text{Br})$ and $\lambda_{\min}$. It is useful therefore to know the range of ϕ angles for which the reflections of a plane with spacing d are in the best condition for recording. This range extends from $0 < \phi < \phi(\text{Br})$, where

$$\phi(\text{Br}) = \text{arc cos} \left(\frac{0.918 \times 10^{-8} \quad (\text{cm})}{2d/n \quad (\text{cm})} \right). \tag{12a}$$

The complete spectrum of x-ray wavelengths also includes the characteristic radiation of the target. This component is very intense in relation to the continuous spectrum. For convenience let us consider only the $K\alpha$ component. In this case (10) become

$$\phi(K\alpha) = \text{arc cos} \frac{n\lambda(K\alpha)}{2d}. \tag{12b}$$

The plane of spacing d has selected this particular wavelength for producing reflections in a particular order n. These reflections are easily recognizable in Laue photographs because of their intensity, as they are produced by the characteristic radiation; these reflections are the Bragg spots of the Laue photograph. These spots, however, have no unusual significance in the photograph, although they do have higher intensity.

Number of orders of reflection in a Laue spot

It is obvious that the total number N of reflections of different orders contained in a Laue spot is the number of lattice points in the corresponding reciprocal-lattice row inside the limiting sphere. In general this number is not known. However, it is obvious from Fig. 2 that the length L of the ith lattice row inside the limiting sphere depends on the ϕ_i, namely

$$\cos \phi_i = \frac{L}{2/\lambda}. \tag{13}$$

The translation period of this lattice row is $1/d$. Consequently, length L divided by the period $1/d$ is

$$N' = L \left/ \frac{1}{a} \right. = \frac{2 \cos \phi_i}{\lambda_{\min}} \left/ \frac{1}{d} \right. . \tag{14}$$

The integer part of N' is the number N of orders of reflection possible for such a reciprocal-lattice row. It is evident that a maximum number N' is reached for $\phi_1 = 0$, in which case

$$N'_{\max} = \frac{2}{d\lambda_{\min}}. \tag{15}$$

This gives the maximum number of orders of reflection that the Laue method can produce for a given rational plane (hkl) of spacing d.

Temperature factor and zone development

The graphical interpretation given earlier provides a straightforward way for understanding the effect that the temperature factor has in the Laue photograph. It is evident from Fig. 3, Chapter 12, that the temperature factor has a decisive effect on zone development. As the temperature decreases, the number of Laue reflections increases in the photograph. The relationship existing between temperature and the Laue photograph can easily be demonstrated with the help of our graphical construction.

As is well known, the temperature factor T enters in the expression of the intensity I_h of the reflection h as follows.

$$I_h \sim |\,F_h\,|^2\,T = f_0{}^2 \exp[-2B(\sin^2\theta)/\lambda^2] = f_0{}^2 \exp(-\tfrac{1}{2}B\,|\,\mathbf{r}^*\,|^2)$$

$$= f_0{}^2 \exp(-4\pi^2\bar{u}^2\,|\,\mathbf{r}^*\,|^2) \tag{16}$$

where $\mathbf{r}^*$ is the reciprocal vector corresponding to the reciprocal-lattice point whose structure factor is F_h, and $\bar{u}^2$ is the mean square amplitude of vibration of the atom; B is the Debye temperature coefficient.

In (16) it is assumed that the crystal contains only one atom per unit cell and that B is isotropic. This simplified example will serve for our purpose. The last expression of (16) indicates that the modification of intensity due to temperature is a function of $|\,\mathbf{r}^*\,|^2$ alone, and that consequently it is independent of the wavelength utilized. This is particularly helpful in the Laue method, for a variety of wavelengths enter in the experiment.

The decrease in intensity I is doubly penalized, for f_0 also decreases with $|\,\mathbf{r}^*\,|$. Furthermore, a reflection can only be observed if its intensity is higher than a certain threshold determined by experimental conditions. Consequently in all practical cases, there exists an $|\,\mathbf{r}^*_{\lim}\,|$ beyond which the reflections cannot be further observed. In the case of the same substance analyzed at different temperatures, $|\,\mathbf{r}^*_{\lim}\,|$ is determined by the value of B (or $\bar{u}^2$). Let us set the limit for observation as 0.01 of the original intensity

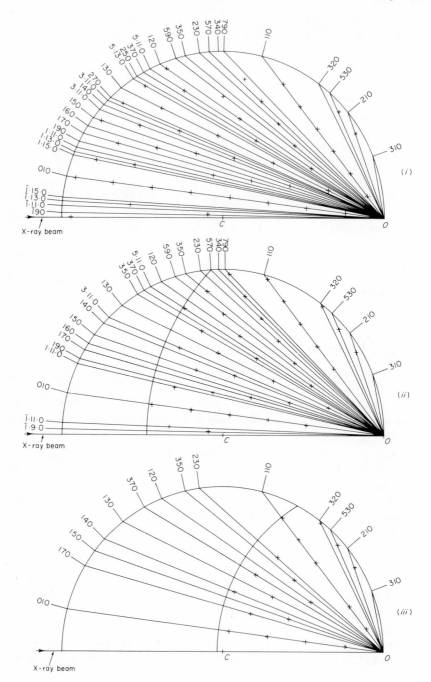

with no temperature effect. Then

$$0.01 = \exp(-\tfrac{1}{2}B \mid \mathbf{r}^*_{\text{lim}} \mid^2), \tag{17}$$

$$\log_e 0.01 = -\tfrac{1}{2}B \mid \mathbf{r}^*_{\text{lim}} \mid^2, \tag{18}$$

from which finally

$$9.2 = \mid \mathbf{r}^*_{\text{lim}} \mid^2 \tag{19}$$

is obtained.

In real cases, consequently, only the reciprocal-lattice points contained inside the sphere of radius $\mid \mathbf{r}^*_{\text{lim}} \mid$ will be able to produce observable reflections. This affects both rotating-crystal photographs and Laue photographs. In the rotating-crystal and powder methods, the reflections are recorded as a function of the angle θ. Accordingly, the temperature effect can be observed by the decrease in intensity of the reflections with the increasing Bragg angle. In other words, as there is direct correspondence between the angle of reflection and film distance from the impact of x rays, a simple inspection of the rotating-crystal photograph gives directly a qualitative indication of the importance of the temperature factor. In the Laue photograph, however, such a straightforward relation does not exist. An illustration of an ideal case is given in Fig. 3, which shows the case of the zone [100] for a body-centered cubic crystal. The incident beam is assumed some degrees off the [010] direction. In Fig. 3, three different temperature factors have been assumed, each one determining a given $\mid \mathbf{r}^*_{\text{lim}} \mid$. In the first case, Fig. 3i, $\mid \mathbf{r}^*_{\text{lim}} \mid$ is larger than $1/\lambda_{\text{min}}$; in the two other cases, Figs. 3ii and 3iii, $\mid \mathbf{r}^*_{\text{lim}} \mid$ is smaller than $1/\lambda_{\text{min}}$. The effect of $\mid \mathbf{r}^*_{\text{lim}} \mid$ is clearly shown in the number of reflections and in the complication of the indices of the crystallographic planes giving rise to the reflections. From those figures it follows that there is an inverse relation between zone development and the temperature factor in Laue photographs. The figures also show that this effect is only visible in the region of back reflection and does not affect the region of transmission (if the actual variation of the intensities is disregarded).

Fig. 3

Author index

Numbers in italics indicate pages on which the complete reference to an author's work is given. Numbers in parentheses are reference numbers and are inserted to enable the reader to locate those cross references where the author's name does not appear at the point of reference in the text.

A

Adams, Oscar S., *53*
Ahmed, M. Saleh, 284, *291*
Amoros, José Luis (J. L.), 199, 202, 224, *237*, 294 (7), 295 (6, 7), 298 (7), 316, 317, 321, *328*
Arguello, C., 104 (28), *121*
Azároff, Leonid V., 211, *238*

B

Baltzer, O. J., 118, *121*
Barker, T. V., 39, 40, 41, 47, *53*, 75 (14), 76, *81*, 140, *147*
Barnes, W. H., *121*, *137*
Barraud, Jean, 332, 333, 336, 338, 342(10), 343 (10), 344, 345, 346, 347, 348, *349*
Barrett, Charles S., *53*, 332, 343 (4), *349*
Baumhauer, H., *182*
Bernal, J. D., 313, 315, *328*
Bernalte, A., 209, 210, *238*
Boas, W., 104 (5), *120*
Boeke, H. E., *53*, *80*
Boldyrew (Boldyrev), A. K., 140, *147*
Born, Max, *349*
Bozman, W. R., 146 (25), *148*
Bozorth, Richard M., *291*
Bragg, W. L., 330, 335, *349*
Brown, Stimson J., *53*
Buerger, M. J., 59, *81*, 182
Burns, D. M., 316 (8), 322 (8), 324 (8), 325 (8), *328*

C

Canut (de Amoros), M. L. (Marisa), 190 (23), 199, 200, 202, 221, 224, *237*, *238*, 298, 316, 317, 321, *328*
Cesáro, G., *265*
Chasles, M[ichel], 264, *265*
Chrobak, Ludwik, 104 (6), *120*
Clark, G. L., *120*
Cloizeaux, A. Des., *52*
Codd, L. W., *53*, 74, 75 (20), *81*, 89, *93*
Cotter, Charles H., *53*
Courant, Richard, *266*
Cox, E. G., 146 (24), *148*
Cullity, B. D., *121*, 210, 212, *237*
Cunningham, R. L., *54*, *81*

D

Decker, Beulah Field, *121*
Deetz, Charles H., *53*
Dijkstra, D. W., *81*
Doliwo-Dobrowolsky, W. W., 140 (12, 15, 18), *147*
Donnay, Gabrielle, 146 (23, 24), *148*, 281, *291*
Donnay, J. D. H., 140, 142, 146, *147*, *148*, *266*, *281*, *291*
Dunn, C. G., 210, *237*, 282, *291*
Dyon, D. J., *53*, *54*

E

Eves, Howard, *266*
Ewald, P. P., 5 (4), *15*

361

Subject index

A

Abnormal zone, 153
Absent reflections, 118
Absorption, 329, 330
Absorption edge
  of bromide, 96
  of silver, 96
Achromatism, 344
Acid, melonic, 161
Acridine, 298
Advantages of stereognomonic projection,
  89
Aids
  in using gnomonic projection, 72–74
  in using stereographic projection, 36–43
Analysis, crystallochemical, 119
Angerer, E. von, 1
Angle
  auxiliary, 177, 179
  axial, 155, 177
  interaxial, 146, 177
  interfacial, 139
  interpolar, 171
  main, 156, 160, 165, 167, 169
  between two poles, 33, 67, 251–252
  between two zone circles, 34
  between two zones, 70
Angle point, of a zone, 69–71, 79
Angle-true property of stereographic projection, 24, 41, 50
Angular data, 142
Anharmonic ratio, 239, 244, 245, 250, 263
Anthracene, 298

Antimony, 282
Apatite,
  form development of, 90
  Laue photograph of, 134
  stereognomonic projection of, 90
Aperture, 332, 338, 343
  circular, 331, 336, 339, 347
  radius of, 347
Apparatus for Laue method, 96, 97
Application
  of cross ratio in crystallography, 250–263
  of Laue method, chief, 119
  to study of thermal motion, 117–118
Area
  blank, 157
  blind, 173
Arsenopyrite, 143
Astigmatism, 345
Augmenting operation, 126, 127
Autocollimator, 64, 65
Auxiliary angle, 117, 179
Axenite, Laue photograph of, 130
Axial angle, 155, 177
Axial photographs, 110–113
Axial plane, 151, 154–156, 160, 161, 180
Axial ratio, 142, 143, 146, 175, 177–179,
  180, 181, 276
Axial row, 155
Axial symmetry, 125
Axial zone, 157, 160
Axis
  crystal, 149, 150, 160, 171
  length of, 146
  rational, 110